Zefs Guide to Deep Learning

Roy Keyes

ISBN 979-8-9873182-0-1

Version 1.0.0

Digital flashcards that accompany Zefs Guide to Deep Learning, as well as the ebook version of the book, can be purchased separately at **zefsguides.com**.

Roy Keyes has been working in machine learning and data science since 2012. He has built and led teams at multiple tech startups and consulted for clients across a wide range of industries. Prior to data science, he received a PhD in computational physics, focusing on medical applications.

His first book was "Hiring Data Scientists and Machine Learning Engineers", a practical guide to hiring data roles. Currently available in paperback and ebook at dshiring.com.

You can find his website and blog at roycoding.com and follow him on Twitter at @roycoding.

Contents

Acknowledgments

Like all books, this one would not exist without the support, input, and assistance of a number of people.

I would like to thank all the people who contributed to this book in any number of ways: I.P., C.A., V.B., J.G., J.J., V.B., J.J., D.T.S., R.M., P.B., A.C., B.P., B.L., J.K., W.C., T.H., F.M., P.J., J.A., D.A., N.S., and some people that I've surely forgotten (you know who you are).

I would like to specially thank early readers that offered constructive feedback and a couple who have opted to have their names mentioned here: Balamurugan Periyasamy and B.S.

This book definitely stands on the shoulders of giants who have made great educational materials about machine learning and deep learning, often freely available to all. Instead of listing all of those creators here, I invite you to check out the links and references found throughout the book.

Big thanks go to all of the people who purchased the pre-release version of this book (and the flash cards, available at zefsguides.com) and those who have supported the Zefs Guides project on social media.

And finally and most importantly, I would like to thank you, the reader, for purchasing this book. I hope you enjoy it and find it useful.

Thank you!

Roy Keyes

Chapter 1

Introduction

This book is about deep learning, a set of machine learning methods that have sparked a huge amount of interest in applying computational and predictive models to everything from whimsical face filters to medical imaging to generating computer code itself. Deep learning is at the core of the current "AI revolution". While based on techniques that can be traced back more than half a century, only in the past decade have these techniques really come into their own and they now dominate the predictive modeling space for an increasingly large number of use cases.

This book aims to help you get a better high-level, conceptual understanding of how deep learning works, its central concepts, applications, limitations, and possibilities.

Why deep learning?

Deep learning is a name applied to a class of neural networks with many "layers", allowing them to be trained to perform certain kinds of tasks that traditional modeling techniques have not been able to do nearly as well. In some cases these deep neural networks can even outperform humans on these tasks, which has fueled the high level of excitement around this family of methods.

Only in the last decade has the capability and potential of these models been truly demonstrated, kicking off a frenzy of interest and subsequent research, development, and usage across industry, academia, and governments. While deep learning has lead to many practical applications that were not reasonably achievable with older techniques, many people also see deep learning as a key component, if not *the* central technology, that will lead to the development of artificial general intelligence (AGI). AGI has been a dream for generations of computer scientists and others going back decades. Where we are on the road to the development of AGI is highly debated, but the practical applications of deep learning are here today and providing value via many real-world use cases.

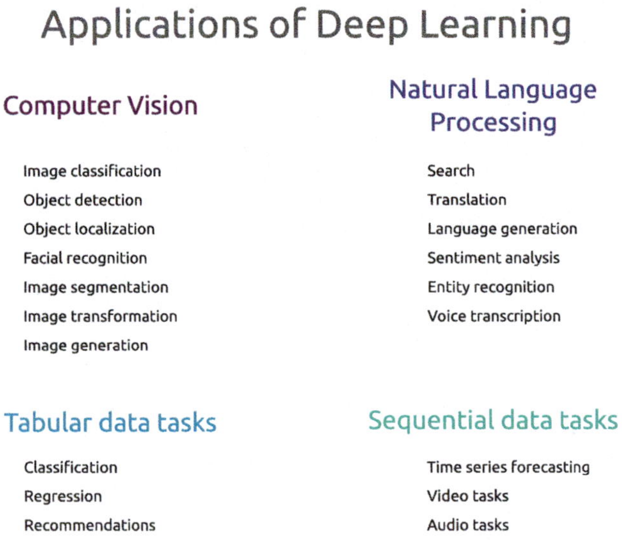

Applications of deep learning

Why this book?

This book is not an academic textbook about deep learning. It will not teach you to mathematically derive backpropagation, implement a GAN from scratch, or how to deploy a vision model "on the edge". This is an intentionally short book aimed at helping you understand the core concepts of deep learning as well as an understanding of where we are today.

If your goal is to prepare for job interviews, do better in your classes, refresh your knowledge for your own work, or have a strong overview of how your team and organization can use deep learning, this book is for you. It is short on purpose to help you get up to speed quickly. Most concepts are presented with accompanying illustrations to better help you understand how things fit together.

What does this book cover and not cover?

This book covers the concepts behind neural networks and deep learning, the key ideas you need to understand to build deep neural networks for solving problems, the common network architectures, and the common use cases that are solved with deep learning. It is, however, not exhaustive. Zefs Guide to Deep Learning is the first in a series of books on Deep Learning topics from Zefs Guides. The current plan for the series includes the titles Zefs Guide to Computer Vision, Zefs Guide to Natural Language Processing, and Zefs Guide to Transformers. Those books all build on the concepts contained in this book.

This book does not cover everything you might want to know about deep learning. It does not go into the detailed math, though I will use a few equations. It does not go into the details of how to implement the training algorithms or model architectures in computer code. Nor does it cover how to use the various programming frameworks, such as PyTorch, Tensorflow, or JAX. It also doesn't go deep into how to solve specific problems, but rather focuses on the more generic, common techniques and problem classes that you can solve with deep learning. Despite these limitations, I think you will find that this book provides you with a quick way to improve your conceptual understanding of the core ideas and topics of deep learning.

How to use this book

This book is designed to help you build up a strong conceptual understanding of deep neural networks, starting with the most basic building blocks of neural networks and moving to use case specific model architectures that are in wide usage right now (2022). What that means is that the first half of the book forms the basis for the rest and should be read through in order to best understand the topics. Chapter 2 is about general machine learning principles. If you're already familiar with "traditional" machine learning you can safely skip that chapter. The second half gets more into specific topics, such as computer vision, natural language processing, and generative models. If you are not interested in one or more of those topics, feel free to skip them.

Available at zefsguides.com[1], but not included with this book, are flash cards that cover the same topics. These flash cards can help you review and recall the concepts in this book, especially when used with a spaced repetition[2] method, such as that employed by the app Anki[3].

If you are interested in going deeper into computer vision, natural language processing, or better understanding transformers, I recommend that after finishing the relevant sections of this book that you continue with one of the other books in the Zefs Guides

---

[1] `https://zefsguides.com`

[2] `https://en.wikipedia.org/wiki/Spaced_repetition`

[3] `https://apps.ankiweb.net/`

series on those topics.

I hope you find this book enjoyable and rewarding to read and wish you the best in learning about these incredible technologies.

Chapter 2

Machine Learning

Deep learning is a family of techniques for building predictive models that are at the center of the current boom in "artificial intelligence". In the past decade deep learning models have been able to surpass "traditional" techniques in many domains and applications, sometimes even exceeding human performance. But to understand deep learning, we need to put it in the larger context of machine learning, as deep learning is itself one type of machine learning. We will also look at shallow neural networks, the ancestors of deep learning.

In this chapter and the next we'll go over some of the core concepts of machine learning and the basics of neural networks, setting us up to learn about modern, deep neural networks.

What is machine learning?

Machine learning (ML) is an approach to solving problems, where data is used directly to adjust the internal parameters of a computer program to provide the best answers possible. Examples include predicting the ultimate sale price of a house or detecting the presence of a tumor in a medical scan. The difference between machine learning programs (usually called "models") and traditional computer programs is that instead of having programmers explicitly write the logic of the program by hand, the important decision logic is "learned" by the program by looking at example data. This process is called training the model.

Machine learning is closely related to (and sometimes identical to) statistical modeling. The main differences are in the historical development and the emphasis on outcomes and explanatory power. Machine learning tends to focus on producing the most highly predictive models, even if they are effectively "black boxes"[1] to the users. Statistical modeling has more emphasis on creating models that have more explanatory power (e.g. creating a model that mimics the underlying process that generates the data). Statistics has also built out robust techniques to deal with small amounts of data, which many practitioners of ML tend to avoid. That said, there are many concepts from statistics that underlie ML and deep learning.

As more data has become available in many areas of science, industry, and government, machine learning has been increasingly adopted as a viable solution to solve real-world problems. The abundance of data is however not the only factor. More data plus cheaper storage and computing power has lead to the development of more software (especially open source software) and the refinement of techniques to make the best use of that data. This has been especially true of deep learning, as we will see in later chapters.

Artificial Intelligence

Artificial intelligence, or AI, is often discussed in relation to deep learning. In fact, many people are actually referring to deep learning these days when the say "AI". Worse, some people even use "AI"

[1]A black box in this context is a process, model, or algorithm where the user is unable to examine the internal logic and can only see the input and output.

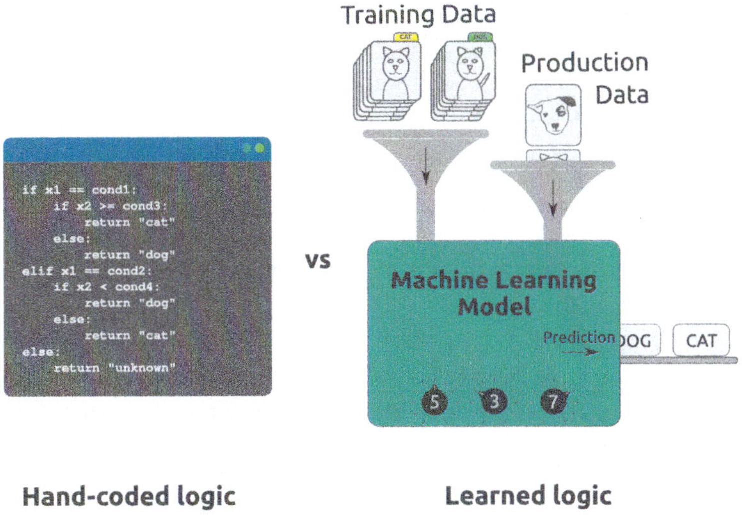

Hand-coded vs learned decision logic

to refer to anything related to data. In this book I will try to avoid the term AI, as it is currently used in a very broad and often vague sense, in part due to the hype surrounding it.

AI has a very interesting history going back decades. It's broadly defined as enabling computers to do tasks that we typically think of as needing some level of intelligence to achieve, whether that be playing chess or holding a conversation. I will not go into the details of this history, except to mention that the two main approaches to achieving this have been so-called "knowledge-based" approaches, with hard-coded logic rules, and "connectionist" or learning-based approaches, which ultimately include the current deep learning techniques.

I am of the opinion that for most practitioners of deep learning, the near-term task-oriented applications of deep learning are where they will get their biggest returns on learning. That said there is a lot of interesting research on how these current techniques might achieve broader "artificial general intelligence" type systems.

For more on AI and AGI, a good place to start is Wikipedia's article on artificial intelligence[a].

[a]https://en.wikipedia.org/wiki/Artificial_intelligence

Types of machine learning tasks and solutions

Most problems that machine learning is used to solve can be thought of as predictions problems. Some fit more into the casual usage of the term "prediction", as in predicting the ultimate sale price of a house, which will only be known at some point in the future. Others are predictions in a technical sense, such as predicting whether a certain image is of a dog or a cat.

Regression

Machine learning tasks can be broken down into a handful of categories. A common way to categorize tasks is by the type of output that a machine learning model needs to produce. One of the most common types of task is regression[2]. Regression is the task of predicting a (continuous) numerical value, such as the temperature in a weather forecast, the sales volume of a product, or estimating time of arrival of a vehicle.

The most basic version of regression that many people are familiar with is finding the slope and intercept of a "best fit line" that fits some data points. This line can then be used to predict other values by using the basic line equation to estimate new, unseen values. This is the manual version of finding the best fit slope and intercept values (a.k.a. parameters). In ML, as we'll discuss in more detail,

[2]In statistics, especially, "regression" is often used as short-hand to specifically refer to linear regression. Here we are talking about it in the broader sense.

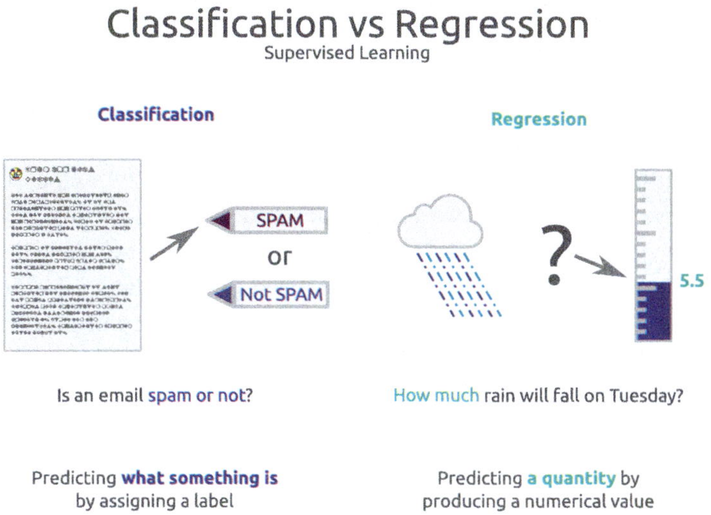

Classification vs Regression

those slope and intercept parameters are determined by the training process, where different values are tried iteratively and the fit of the line is compared with the known data to see how good the fit is. This is the core of machine learning: "learning" the best parameter values of a model from the known, existing data.

Regression is one of the most common applications of machine learning and typically falls under the category of supervised learning, which we will discuss below.

Classification

Classification is the task of predicting the class, category, or label of an item. This can be a task with many possible choices, such as facial recognition, or a binary, yes/no scenario, such as predicting whether a customer will purchase an advertized item. Other examples include predicting the species of an animal present in an image, the sentiment of a product review, whether a self-driving car needs to make an emergency stop, and how a support request should be routed.

A simple classification algorithm is the flow chart, which is similar to the decision tree. A medical doctor might create a flow chart to map out how to make a certain medical diagnosis, based on symptoms, medical history, and measured patient data. The doctor does this based on their knowledge and understanding of how they make decisions based on the available data. The doctor creates a decision criteria for each piece of input data, such as body temperature. The path of the flow chart, based on each decision, leads to the medical diagnosis, or "classification". A similar model can be created with machine learning. Instead of the doctor manually inputting the decision thresholds and the overall flow of the decisions, the decision criteria are learned from the available data by trying out many combinations of decision thresholds and structures and evaluating the overall goodness of the predictions, until the best decision thresholds and flow structure is found.

As with regression, classification tasks typically fall under the umbrella of supervised learning. Sometimes you don't know what the classes are for a classification problem, so instead you try to discover clusters or grouping in the data based on similarities. This is unsupervised learning, which we will discuss below.

Supervised learning

Another way to organize machine learning tasks is by the method used to solve the problem, rather than the type of predictive task. One of the most common classes of methods used to build predictive models is "supervised learning". Supervised learning is when you have data to train a model that includes "ground truth" answers. For example, if you were building a model to predict how

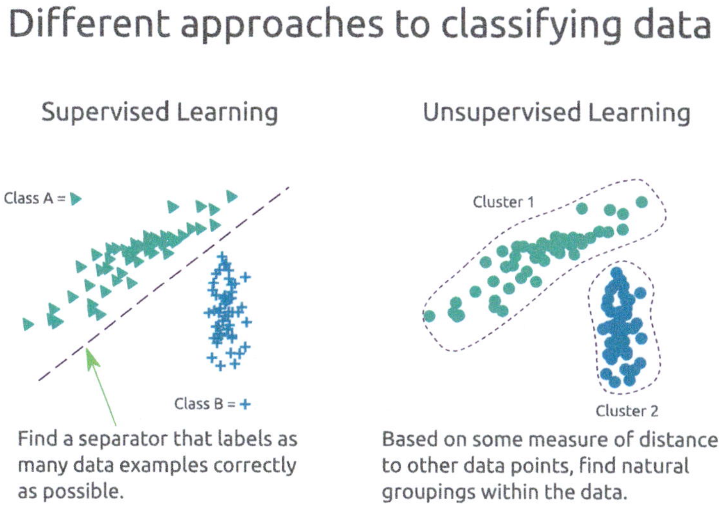

Supervised classification vs Unsupervised classification

tall children would be as adults, you could use a supervised learning approach if you had data on full-grown adults when they were children and their final, adult heights. The inputs to the model would be the earlier data, such as age, height, height of parents, etc, and the matching outputs would be the known, final heights of those people.

Essentially supervised learning is when you train a model by taking example inputs and comparing the predictions of the model to the known, desired outputs. By altering the parameters of the model (think the slope and intercept of your fit line), you can then check if the new parameters provide an improvement in the overall predictive performance of the model. In practice this is done by training algorithms, which automatically try many possible model parameter values to land on the best model, given the current data.

Unsupervised learning

In contrast to supervised learning is "unsupervised learning", where you do *not* have the "ground truth" answers for the task you are trying to address. One of the most common tasks using unsupervised learning is "clustering". When clustering, you want to discover natural groupings of data, based on similarities within the data set. For example, you might want to cluster customers by purchasing behavior or demographic information to create focused advertising campaigns. You may not know what these groupings are beforehand or how many groups to expect. Only after investigating the members of these clusters can you (potentially) give them labels that make sense. This is in contrast to supervised learning, where you know beforehand how many classes you have and what they are. These additional challenges can make unsupervised learning more difficult to perform, but it may be worth it if collecting the "ground truth" is very costly or even impossible.

Self-supervised learning

There are some approaches to solving problems that use "unlabeled" data, but in a different way than the unsupervised approaches mentioned above. Instead of searching for natural clusters or patterns within the data, you can pose tasks where the answer is already contained in the data. Examples include training models to predict the next word in a sentence, using the first part of the sentence as the input and the following word(s) as the "ground truth" output, effectively formulating the task as a supervised learning problem. This reformulation from "unlabeled" data to a supervised learning problem is why the moniker "self-supervised" is used, as the data is its own label.

The distinction between supervised, unsupervised, and semi-supervised learning is not always clear-cut, but it is often a useful one[3].

Reinforcement learning

Another category of machine learning is "reinforcement learning". Reinforcement learning is an approach to solving tasks where a strategy for decision making is learned, typically trying to maximize or minimize some score or ultimate goal. Common examples of tasks that reinforcement learning is applied to are games, where the player needs to make a series of decisions, given the state of play. The reinforcement learning model may learn from the score in the game, winning or losing, or all of those combined. Reinforcement learning combined with deep learning has been used successfully in many gaming tasks in recent years[4]. Reinforcement learning will not be further covered in this book.

An example task

Before we look at the details of how machine learning problems are formulated and solutions are built and tested, let's consider an example problem. This will give us even more context when we discuss the specifics of how machine learning solutions are created.

Predicting real estate sales prices

Let's imagine that you decide that you want to buy some real estate (e.g. a house). You want to use your computing skills to give you the best understanding possible of how much to pay for a house by building a program that could predict how much a given house would ultimately sell for. How would you do this?

---

[3]There is another category termed "semi-supervised" learning that we will not cover in this book.

[4]`https://en.wikipedia.org/wiki/MuZero`

As we have seen above, this is a regression problem. We want to know a specific numerical estimate: the price in some currency. If you were a physicist, *cough*, you might try to build a model from first principles. One based on assumptions about how humans act and information about the fundamental attributes of the house. This probably won't work, as the underlying mechanisms are far too complicated. The machine learning approach would be to try to use data to adjust a mathematical model, such that the model can effectively use the different characteristics of the house to predict the sales price. By collecting data from recent sales of homes in the region, you could use the known characteristics, or features, of these homes as training inputs to an ML model and compare the predicted sales prices to the known, actual sales prices.

Assuming you had selected a model and trained it using the data you had, how would you know if the model was good? A good model should produce price predictions that are close to the actual sales prices of homes in the area. While training, you would compare the outputs of your model to the known values, adjust the internal parameters of the model, check the overall goodness of the predictions, and iterate this process until the model is no longer improving or you are satisfied. This training process is done via a training algorithm, which is usually specific to the type of machine learning model you are training.

One immediate issue with this is that your model may be able to learn to predict the sales prices of the examples you have by essentially memorizing them, but, when faced with new data, not be able to make reasonable predictions. You model has focused on memorizing the training data, rather than on learning the more general patterns that are useful for predicting sales prices of houses that it has not yet seen. The way to understand and quantify the goodness of your model is to test it on house sales data that it did not see during the training process. This will help estimate how it will do with "real world" data.

The basic steps in this example are:

1. Formulate the problem
2. Collect necessary data
3. Train the model
4. Evaluate the performance of the model
5. Iterate as necessary

There are a lot of details they I have glossed over in this example of a supervised regression problem, but we have now set the stage to dive into those details more deeply.

Formulating machine learning problems

Machine learning is not the solution to all problems[5]. Some problems are well suited to a machine learning solution, but many are not. You must first decide if ML is the right (or reasonable) approach to solving the problem at hand. If a (simpler) non-ML solution makes sense, go with that. If ML makes sense, the next thing you need to do is formulate the problem in a way that ML can be applied.

Formulating the problem for ML means identifying what kind of problem it is and the general approaches that make sense. In the previous example, we recognized that the problem of predicting the ultimate sales price of a house could be posed as a supervised learning regression problem. If we knew the characteristic "features" of houses in a statistically representative set that had sold in the past, along with the sales prices, we could try to train a regression model to predict sales prices of other houses. The key to this is recognizing the problem and what data is needed.

Often times a problem cannot be approached directly or lacks the data needed. If you want to predict the preferences of new

[5]This bears repeating: ML is not the answer to all problems!

visitors to your website, but know nothing about them, you are unlikely to be able to predict much beyond the general trends of your visitors as a whole. If you have an extensive history of what a specific user likes, you are in a much better position to make a specific prediction.

Another important aspect of formulating machine learning problems is landing on exactly *what* you are trying to predict. If you are trying to maximize the lifetime value of customers, for example, it's very important to make sure that you understand how a customer generates value for your business. It might be something as obvious as the total amount of money they spend on your products over time, but it might instead be about how many times they interact with your content if your business is centered on advertising. Ultimately this is driven by your business model (or organizational goals and strategy), but when it comes to ML, the problem needs to be correctly distilled into an specific, actionable prediction task.

Data sets and features

Machine learning is truly "data-driven". ML models learn via training data, whether supervised or unsupervised. Without enough high-quality data for your task and model, you will not get the results you need. ML practitioners are often faced with the problem of not enough and/or poor quality data. If, for example, you are trying to predict the life span of ships, but only have data on cruise ships, you are unlikely to do well predicting the life span of fishing vessels. If you have a ton of data, but it's missing key fields, or data about the same events cannot be tied together, your data may be too "dirty" to allow you to achieve what you want.

The data fed into ML models contains so-called "features". Features are the characteristic aspects of the events or entities in the data. Examples of features related to the real-estate example above would be the size, age, location, style, and condition of the house for sale. Supervised ML models try to use these features to make predictions by finding correlations between the features and the predicted quantity or label (often called a target variable). Unsupervised ML models use these features to find similarities between examples in the data.

Much of what the ML practitioner spends time on is gathering, cleaning, and formatting data to make it useable for modeling. Additionally much time and effort is spent on making sure that the most predictive features are present. Sometimes a raw feature, such as the size of a house, is highly predictive, but other times features must be "engineered" for the most predictive ability. If you were trying to predict disease risk of some disease, height and weight might be less predictive that a composite quantity, or feature, such as body mass index. There are many techniques to create these more abstract features and ML practitioners commonly try many of them to see if they will improve the predictive performance of their models.

Finally, it's crucial that the data you are using is truly representative of the "real world" data that your system is intended to work with. This can be difficult to assess, as there are many ways that data can be non-representative, such as biased sampling, snapshotting data from a changing system, or even collecting data from the wrong sources.

Measuring performance

Probably the most important question about a machine learning model is how good the model is. This question sounds straightforward, but how you measure the performance of a model is something that ML practitioners often must spend a lot of effort on.

For regression tasks, common measures of performance are the mean absolute error, MAE, and the root-mean-square error, RMSE. These quantify the typical error of predictions made by the model.

Both look at the difference between a prediction and the "ground truth", while treating an overestimate the same as and underestimate, but RMSE effectively "focuses" on large errors in a disproportionate way. This is often preferred, if large prediction errors are especially costly. In contrast to this, you might have a task where overestimation is fine, but underestimation is very costly, so using a symmetric metric, such as MAE or RMSE would not be appropriate and a "weighted" approach would be better suited.

Classification tasks have their own set of performance metrics and even more explicit tradeoffs, depending on the issues related to incorrect predictions. Common classification metrics include accuracy, precision, recall, specificity, and F1 score. We will discuss some of these in more detail when we talk about classification tasks for deep learning.

Regardless of the task, the most appropriate performance metric should be chosen, taking into account the pros and cons as relates to the specific task.

Performance baselines and success thresholds

One of the difficulties in trying to solve problems with machine learning is that large amounts of data and long training times are often needed before good performance levels are achieved. The practical drawback of this is that "starting small" and moving incrementally is often not feasible. That means that it's common to dive in with what seems like a good idea, only to discover that the idea does not result in a good model after much effort.

One mitigation strategy is to establish performance baselines. By quantifying the performance of the current solution, you can set a baseline for how well the new model needs to perform. If there is no current solution, you can create a baseline solution by starting with something as simple as possible, such as using the mean price or a very simple linear model for predicting real estate prices. If your model does worse than that simple baseline, you know that you are on the wrong track[6].

Along with performance baselines, it's very important to establish success thresholds for models. How far off can your real estate sales price predictor be and still be considered good enough? $100, $1000, $10,000? This is even more important when you are building models at the request of others who will ultimately rely on those models.

Model selection

Once you have formulated the problem, you need to select a machine learning model and train it. In practice, many ML practioners don't simply select a model and run with it, but rather select a set of models and see which ones perform best for the task at hand.

Generally there are sets of model that are suitable for different types of tasks, which allows ML practitioners to narrow down their choices. The type of problem and amount of data are key factors in choosing a model. Practical considerations related to training resources needed, prediction speed (a.k.a inference speed), explainability, simplicity, deployability, and current infrastructure come into play when deciding on which models to consider, as well.

Boosted trees are all you need?

While it is common to assess a number of different model types when working on an ML problem, sometimes people will stick with what they know best and that seems to work well across different problems. Currently, outside of problems related to computer visions and language, gradient boosted trees is one of the most popular models, especially for tabular or structured data.

To quote a tweet from Kaggle Grandmaster Bojan Tunguz[a]:

[6]Occasionally you'll create a very simple model as a baseline and find that it is actually good enough for what you want to do!

So here is an easy two-step process for guaranteed success:

1. Show up.
2. Use XGBoost.

ML competitions like those on Kaggle have helped demonstrate which machine learning models are robust across different kinds of problems. Implementations of gradient boosted trees, like XGBoost, have proven very successful, though using them as first choice is not without debate.

[a]https://twitter.com/tunguz/status/1455954397687685129

Model training

Training a model is how you go from a set of data and a raw model to something useful. Models have internal parameters that need to be tuned to provide the best predictions related to a task. "Learning" the best parameters is achieved by using the (hopefully high quality) available data.

Supervised learning

For supervised learning problems you have feature data, which describes the example (e.g. a house, picture, website interaction, patient symptoms, etc), often denoted with the variable x, and the so-called target data, which is the outcome, quantity, or label (e.g. sales price, content of the picture, what someone clicked on, diagnosis, etc). The target is often denoted with the variable y.

To train your model you will adjust the model's parameters, such as the slope and intercept terms of a simple linear model, provide the model with example input features, and compare the model output, often labeled $\hat{y}$, with the known "ground truth" target, y. Based on the agreement or error between the predicted target value and the actual target value, you will then adjust the parameters and once again pass in the input features and see how good the predicted outcome values are. This iteration is carried out until the predictions are "good enough" or have stopped improving.

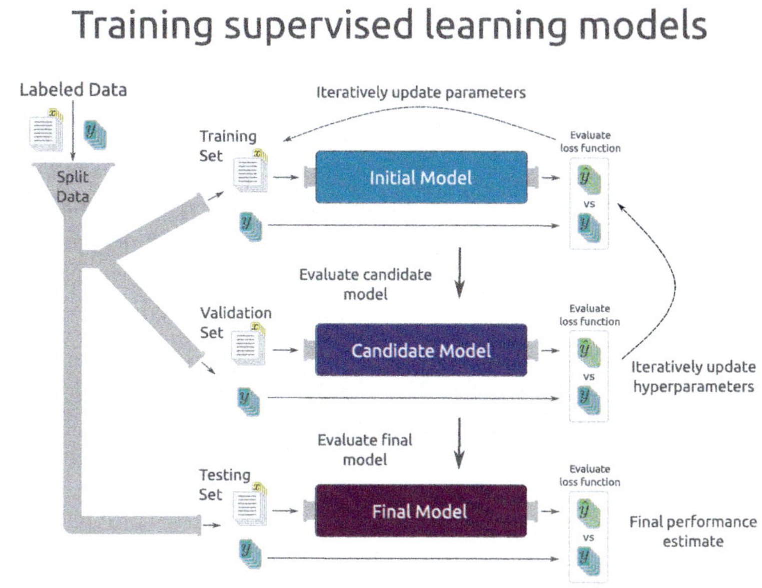

Training supervised learning models

Splitting data sets for training

For training supervised ML models you need a (high quality) labelled data set, but you also need to use the data in a specific way. A supervised model learns by taking in data, x, producing predictions, $\hat{y}$, and comparing those predictions with the "ground truth", y. But to understand how well the model will perform on as-of-yet unseen, "real world" data, another data set that the model has not yet seen is needed.

In order to make this possible, the original data set is first split randomly into a training set and a "test" set, which is only used to estimate the performance of the model once the training is complete. Typically a small fraction between 5% - 25% is held out for the test set. It's very important that these sets are kept separate so that there is no "leakage" between them that would let the model learn the specific examples that are in the test set.

In practice a third set is also split out of the training set after the initial split. This is called the validation set and is used to optimize so-called "hyperparameters". In contrast to model parameters, which are learned directly by training the model, hyperparameters are model settings that are set by hand. Examples of hyperparameters are the number of trees in a random forest, the degree of the polynomial in polynomial regression, and the number of layers in a neural network.

The validation set is used after a model has been trained on the training set to do an initial estimate of the goodness of the model. After that, the hyperparameters are varied and the training process is carried out again to evaluate another set of hyperparameters[7]. Finally, the model with the best hyperparameters is tested against the test set, producing the final model performance estimate.

Unsupervised learning

In unsupervised learning, you have no target or outcome data (a.k.a. data labels). Instead, the model is using input features as the basis for grouping or clustering the data. This can be used to find natural groupings of customers, such as "big spenders" or "browsers", by finding behavioral similarities. This can also be used for finding anomalous and fraudulent behavior that falls outside of the main clusters.

[7]There are several common approaches to hyperparameter search, such as grid search, random search, and various Bayesian methods.

Typical clustering algorithms iterate through the data, trying to place each data example in a cluster that best fits it. Other clustering algorithms try clusters with slightly different parameters until most data examples seem to fit well within a cluster. Most clustering algorithms require that the user set the number of expected clusters that should exist in the data. The optimal number of clusters is not necessarily known beforehand, but there are procedures and metrics, such as the elbow method, the gap score, and the silhouette score, that can help the modeler find the best number of clusters.

The clusters resulting from unsupervised methods do not always correspond to obvious groupings and typically require human inspection to assign meaningful labels (if that's important). For example, if you clustered a large set of songs, you might find that the most important feature was the tempo of the music, which might put otherwise desperate genres together.

Dimensionality reduction

Another type of unsupervised learning is dimensionality reduction. Dimensionality reduction is a way to take many features and reduce them to a smaller number of features, which can be important for working with limited computing resources. This is done by finding new features within the data that are more correlated with other aspects of the data (e.g. the target variables, if those are available) and ranking those, such that the highest ranked new features are the most predictive ones.

One of the most common dimensionality reduction techniques is principal component analysis (PCA). PCA finds a set of new features called "principal components", which best explain the variance in the data. By keeping a smaller number of these principal components than the original number of features, you can reduce the resourced needed for training supervised models and potentially remove some noise in the data.

As with groupings found in clustering methods, it may be difficult to interpret exactly what the principal components represent in the data.

Loss functions

When performing supervised learning, you are comparing the predicted values with the known, "ground truth" target values (i.e. $\hat{y}$ vs y). To quantify how well the model has predicted the target values, a loss function is used[8]. A loss function quantifies the difference between $\hat{y}$ and y across all examples used in training.

For regression tasks, mean squared error (MSE) is a common loss function:

$$L(\theta) = \frac{1}{n}\sum_{i=1}^{n}\left[y(x_i) - \hat{y}(x_i, \theta)\right]^2$$

for some given parameters of the model, denoted by θ, and n training data examples.

This has the advantage of treating overestimates and underestimates equally, while having some other convenient mathematical properties[9]. It also places more weight on larger errors than a simpler, (piecewise) linear measure, such as mean absolute error ($L(\theta) = \frac{1}{n}\sum_{i=1}^{n}|y(x_i) - \hat{y}(x_i, \theta)|$)

Classification tasks predict the class or label of the input and thus need a different type of loss function. The loss function needs to compare the label assigned by the model and the "ground truth" label. There are several common loss functions used for evaluating the performance of a classifier. For binary classification (i.e. only

[8]The term "loss function" and "cost function" are often used interchangeably, though sometimes they are used to mean slightly different things. Here I will use "loss function" as the generic term for measuring how far a model's output is from the desired output.

[9]Specifically the the derivative with respect to the model's parameters is a very simple form.

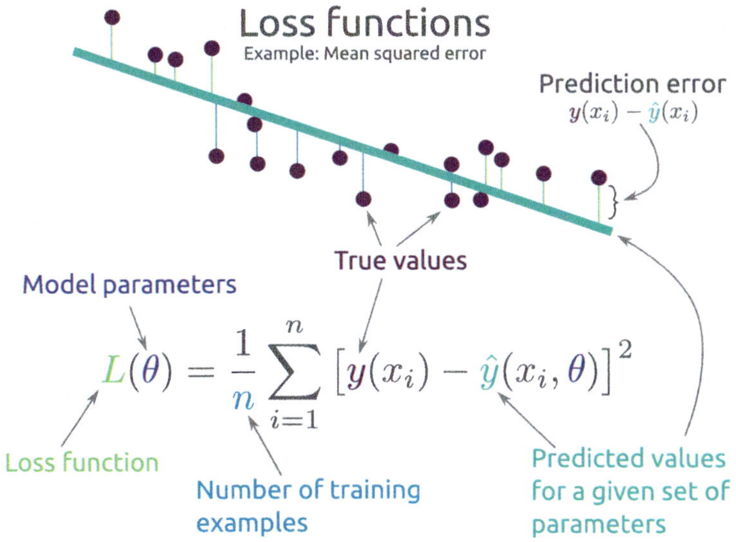

Loss functions: Mean squared error

two possible classes) common loss functions include binary cross-entropy (a.k.a. log loss), hinge loss, and Huber loss. Common loss functions used for multi-class problems include multi-class cross entropy and Kullback Leibler Divergence.

We will look at these more in depth in the sections on deep learning.

Parameter optimization

Finding the best parameters for a model is an optimization problem. We want to know which values result in the best predictions. Machine learning training algorithms iteratively search for the best parameters in various ways. The loss function is the training algorithm's guide to which values to try and it's generally chosen to make finding the best parameters as fast and robust as possible.

Depending on the type of ML model, the optimization make take place globally or locally, i.e. the training algorithm may look at the total error or at some loss related to the specific step in the machine learning model algorithm itself[10]. We will consider global optimization, which is used by many traditional ML model types, but is especially relevant to neural networks.

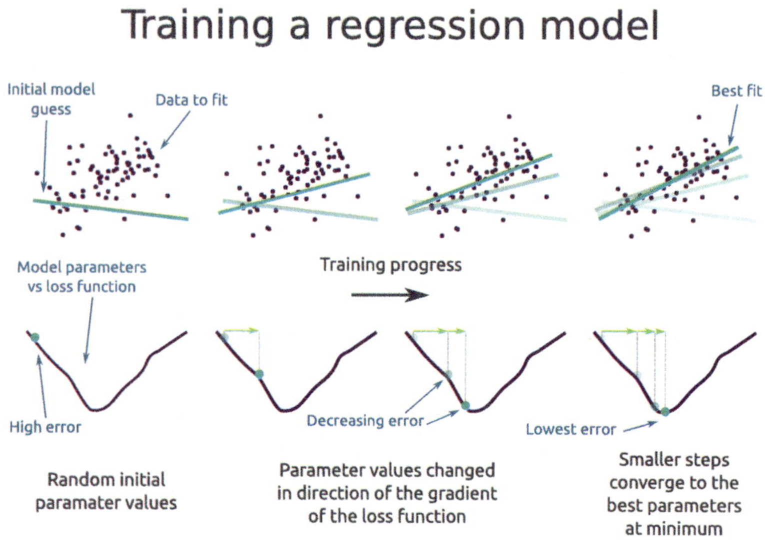

Training a regression model

ML model training can be thought of as searching for the parameters that result in the lowest prediction error. This is often thought of in terms of "parameter space", which is a sort of abstract landscape, with hills and valleys representing high error levels and

[10]Tree-based models, such as decision tress and random forests, optimize decision nodes locally.

low error levels, respectively. Training is the process of searching through this parameter landscape for the deepest valley, or global minimum, which represents the lowest prediction error. It's easy to get fooled by "local minima", which look like the deepest valley, but are actually not as deep. The more complex and "bumpy" the parameter space, the more difficult it is to find the global minimum. The smoother the landscape, the easier it is.

Gradient descent

The method most commonly used to for finding model parameters that minimize the loss function is called gradient descent. Gradient descent is the process of moving "downwards" along the slope, or gradient, of the loss function in the parameter space. The process is to adjust the parameters in the direction of the (negative) gradient step by step, moving "down hill" and re-calculating the gradient after each step. The size of the step is determined by the steepness of the gradient and a step size multiplier, or learning rate, set by the user. Eventually the steps should lead to a local or global minimum, usually after taking smaller and smaller steps. Stopping is determined by some criteria set by the user.

From a mathematical perspective, the gradient of the loss function is found by estimating the derivative of the surface with respect to the parameters, which is the same as the slope, but often on a surface that exists in a very high dimensional space.

There are several different algorithms for performing gradient descent and we will look at a few in more depth in the sections on deep learning.

Generalization and overfitting

The goal of learning from a data set is not for the model to simply memorize the training data, but rather for the model to recognize the patterns in the data, such that it can make good predictions about data from the same distribution that it has not yet seen. In other words, we want out cat image detector to work for all images of cats, not just the ones it's already seen. This process of going from training data to being able to make prediction about as-of-yet unseen data is called "generalization".

The opposite of generalization is called "overfitting". Overfitting is when the model has focused too much on the specific quirks or noise of the training data and has not been able to learn the more general patterns within the data. Overfitting is a common issue in training ML models and is one of the things that practitioners worry about the most.

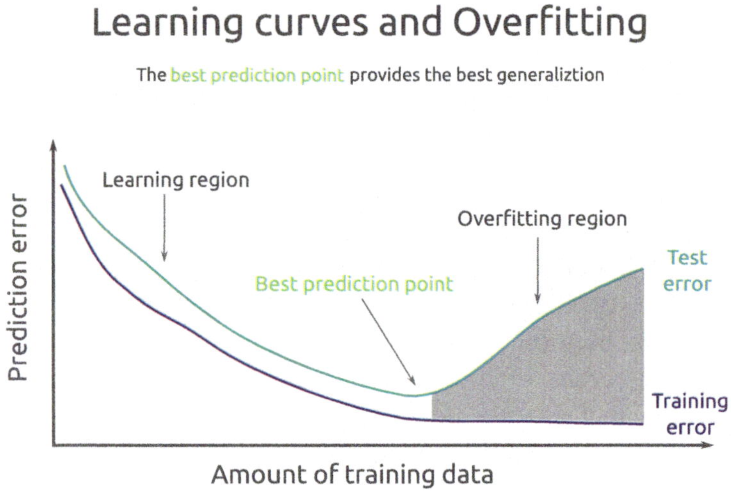

Learning curves, generalization, and overfitting

In order to understand if a model has generalized well, the prediction error can be determined for both the training set and the test set, which the model has not yet seen. Typically as more data is used for training, the performance of the model increases and the prediction error decreases. The overall change in prediction error as the amount of training is varied creates what's called a learning curve. Plotting a learning curve can help you see exactly where you are in the learning process. For example you may be in a region where a lot of improvement is occurring and adding more training data will be the best thing to do. Conversely the model may have plateaued and stopped learning.

By plotting both the prediction error using the training set and the test set, you can see if the model has begun overfitting. The figure above shows a classic scenario of the model improving on both the training set and the test set, but the two learning curves start diverging once the model begins overfitting to the training data. Once it starts to fit too closely to the training set, instead of learning the more general patterns in the data, it will do worse and worse on the test data[11]. The best parameters for the model are the ones from the "sweet spot", where the test error is lowest.

Bias and variance

Two of the important ways to quantify how well a model behaves are bias and variance. In general you can think of bias as how consistently off an estimate is from the true value. More specific to ML, bias is how far off a model trained with a given set of hyperparameters is from the the true target value when being used to predict the target value from a test set. In other contexts this is called systematic error, rather than random error or noise. To characterize the bias, multiple estimates of a value need to be made. The mean error of those estimates is the bias.

For ML models, the bias is the mean error from multiple models all with the same hyperparameters, but trained on different sam-

[11]There is a related phenomenon called "double descent", where the test error may go up and then back down again, but we will not cover that here. See `https://en.wikipedia.org/wiki/Double_descent`.

ples of the training data. This estimates how far off(and in what direction) a specific model will be from the "ground truth".

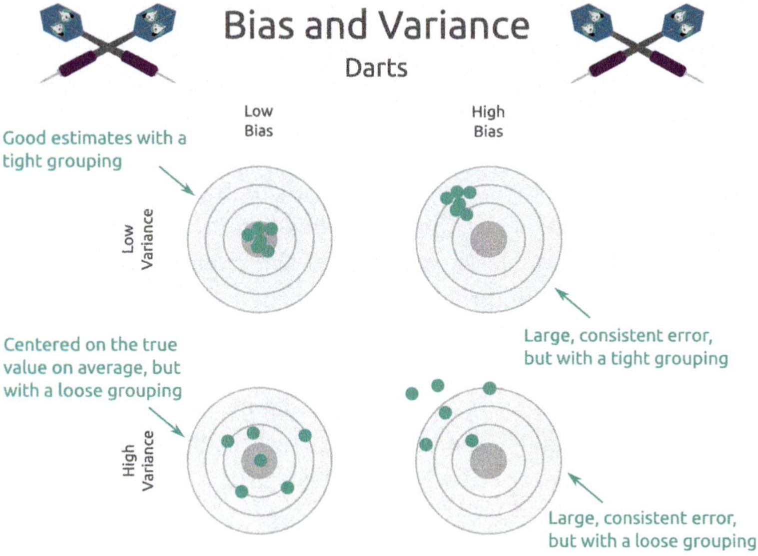

Bias and variance in estimates

Variance is a measure of how "spread out" a set of values are. For ML models this is the spread of predictions for a single point in the test set, as made by several instances of the same model (i.e. same hyperparameters) trained on different samples of the training data. A model that overfits will results in several models that give widely varying predictions of the same input, as each model has learned the very specific patterns in the samples, rather than the more general pattern common to all samples.

The ideal model has low bias and low variance. In practice there is typically a trade off between having either a low bias or low variance model and the best choice is a moderately low bias and moderately low variance model.

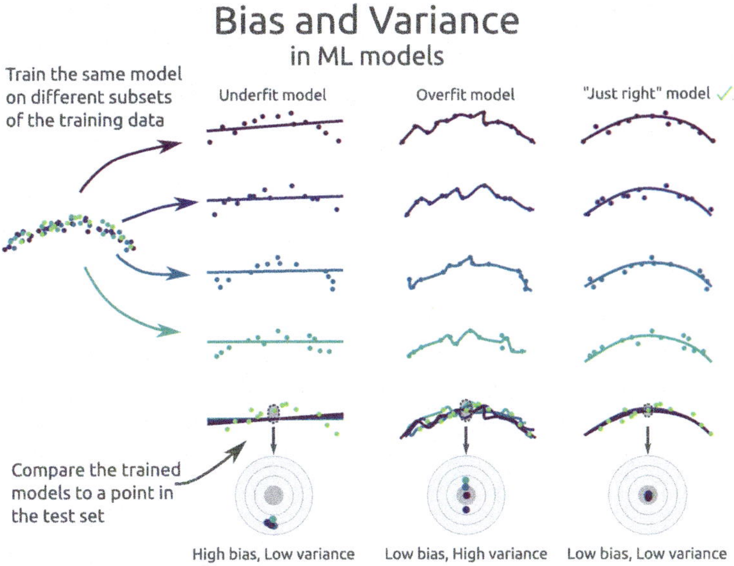

Bias and variance in ML models

Avoiding overfitting

Overfitting is a problem with many kinds of machine learning models and practitioners put a lot of effort into avoiding or mitigating it. There are several strategies for this.

The simplest strategy to avoid overfitting is to use more training data. The more data the model needs to fit well, the less chance there is for it to fit to noise in the data. This is not always possible, as there are often costs to either collecting and labelling more data or computing with more data.

Regularization

Overfitting, high variance models are typically the result of too much flexibility in a model. If a model has enough internal parameters to match every quirk of the data, it's less likely to settle on the "smoother" general patterns in the data and more likely to try to match noise. One way to use a flexible model while avoiding overfitting is by constraining the parameters in the model. This approach is a form of so-called regularization.

Instead of allowing the parameters to take any values they want, regularization of this type constrains the parameters in a way that limits how much weight can be used in the model. If the model parameters are constrained to some total magnitude, then the model is more likely to distribute the parameter values to emphasize the more fundamental patterns in the data, rather than noise or edge cases.

This type of constrained regularization is typically achieved by adding a term to the loss function, which increases the loss function when the sum of the parameter values is high. An example is L2 (or ridge) regularization:

$$L(\theta) = \frac{1}{n}\sum_{i=1}^{n}\left[y(x_i) - \hat{y}(x_i, \theta)\right]^2 + \lambda\sum_{j=1}^{k}\theta_j^2$$

where λ is a hyperparameter that allows you to change how much regularization to use (i.e. larger values of λ require smaller values of the sum of [squared] parameters, which usually results in the more fundamental patterns being emphasized). A λ of zero would mean no regularization.

If you are applying a model such as a set of Legendre polynomials, the parameters are the multiplying coefficients on the different polynomial degrees:

$$\hat{y}(x, \theta) = \theta_0 P_0 + \theta_1 P_1(x) + \theta_2 P_2(x) + \theta_3 P_3(x) + \theta_4 P_4(x) + \cdots + \theta_k P_k(x)$$

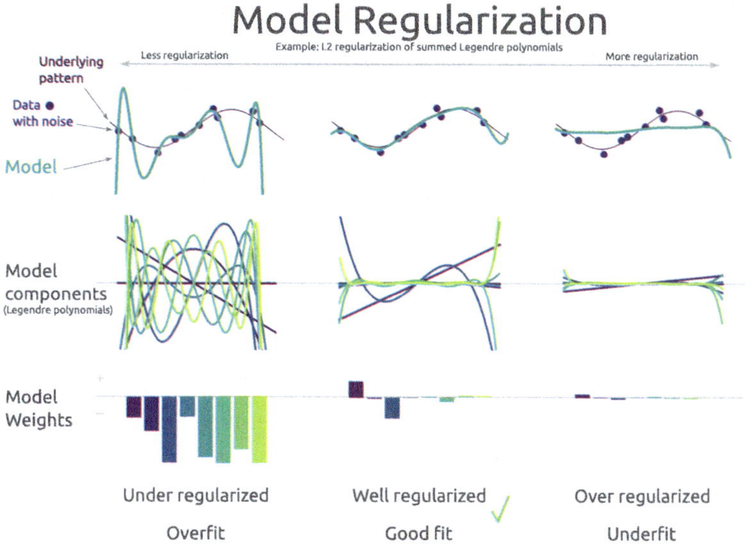

Model regularization (example with Legendre polynomials)

The higher order (i.e. higher index) terms in the Legendre polynomial model are "wigglier", which allows them to capture more small-scale change in the data, which is often due to noise. By constraining the sum of the (squared) parameter values, the lower degree terms are more likely to be emphasized. Conversely, too much weight on the lowest order terms can cause underfitting. A good fit will balance the terms and requires just the right amount of regularization.

There are several other forms of regularization, some of which we will discuss in later chapters. All are designed to create models that are more likely to model the general, underlying pattern of the data and less likely to focus on the noise in the data.

Hyperparameters

Hyperparameters are aspects of a machine learning model that are not directly learned by training on the training data, but are set by the person training the model. Examples of hyperparameters include the number of trees in a random forest, the number of layers in a neural network, and the number of clusters in a clustering algorithm.

When trying to find the best performing model, hyperparameters are one of the many things that can be changed and experimented with. The default parameters for a given machine learning model (from an ML software library) may not produce the best results, therefore machine learning practitioners typically spend a fair amount of effort trying to find the best (or at least "good enough") hyperparameters.

As discussed in the section on data set splitting, hyperparameters are typically evaluated against the validation set. A model's hyperparameters will be changed, then trained on the same training set, and then evaluated on the validation set. While this type of iterative search for the best hyperparameters can be done by hand, it's typically done via a search algorithm.

Common hyperparameter search techniques include:

- **Grid search**: a grid of values (e.g. 0.1, 1, 10, 100, etc) are tried for each hyperparameter and combination of hyperparameters. Grid search has the advantage of being parallelizable, but has the drawback of potentially taking a very long time when there are many hyperparameters (and thus many, many combinations).
- **Random search**: combinations of hyperparameters are randomly chosen from possible hyperparameter values. This has the advantage of being able to evaluate combinations in a very large space of combinations, but still having a high probability of finding good combinations in a reasonable amount

of time.

- **Bayesian search methods**: an iterative technique using Bayesian methods to update which parts of the hyperparameter space the best combinations are likely to be found. This a more complicated set of methods that are not very parallelizable, but can potentially find good hyperparameter combinations more quickly.

Unlike the parameters of most ML models, hyperparameters cannot typically be optimized in the same way that model parameters can be optimized, using techniques like gradient descent. This is because the hyperparameter spaces are typically not smooth and differentiable, but at least partially discreet and discontinuous. For that reason the above methods are applied.

Productionization

Finding the best model through the iterative training process is not usually the end goal of an ML project. The end goal is to actually use that model in some real-world capacity. The model needs to be put "into production" or deployed for real-world use.

Deployment can look very different in different settings and contexts. E.g. creating a model that is run manually once a month is very different from a model that runs continuously on a mobile phone. The most common scenario of deployment for most models is as an API as part of a larger, server-based system, such as a product recommendation system.

While the different productionization scenarios have their own sets of tools, common themes are deployment, monitoring, and maintaining models. It's often said that development of an ML model is never finished, because while a model is usually static, the world is constantly changing. This means that once it's deployed, whether as a web API or on an edge device, the owner of the model needs to keep track of its performance. Some scenarios, such as predicting the weather lend themselves easily to performance monitoring, as the "ground truth" is easily available. Other scenarios, such as detecting infrequent anomalies in long cycle systems, may take time and effort to understand the model performance. Tools and procedures for these may exist or may need to be developed by the engineers.

Maintenance of models is needed when the current model is no longer good enough. That may be because the "world" has changed and the previous predictions are no longer accurate[12], there are new requirements, or a better version of the model has been developed. A key feature of a robust ML deployment system is the ability to track versions of models and to easily roll-back from a model that's not working to one that is. This is where the research and development aspects of machine learning really meet the software engineering aspects of machine learning.

Common issues

Machine learning is a very powerful approach to solve certain kinds of problems, but it's also incredibly easy to get wrong[13]. There are a number of common mistakes that practitioners make and anyone working on ML models needs to be vigilant to avoid these.

Overfitting, described above, is a common problem. Without monitoring training closely, most ML model types can overfit easily. Strategies to mitigate and avoid overfitting include training on more data, regularization, early stopping, and cross-validation (to better estimate performance), among others.

[12]You can imagine the impact that the CoViD-19 pandemic had on many models that were previously running well.

[13]A very good talk about this was given by Ben Hamner from Kaggle in 2014 called [Machine Learning Gremlins](https://www.youtube.com/watch?v=tleeC-KlsKA), laying out many of the ways that ML can go wrong. `https://www.youtube.com/watch?v=tleeC-KlsKA`

Not having enough data can also lead to low performance. This can be diagnosed by looking at the learning curve(s) while training. If the learning curve is still showing improved performance (i.e. decreasing error), despite having trained on all of the data, you are likely to get better performance by training with even more data.

ML models only know what they've learned from their training data. Even if they are correctly trained, if the underlying data is wrong in some way, your model will have problems when being applied to the real-world problem. This can be due to non-representative sampling, which can cause bias in the model's predictions or cause the model to learn non-essential context, rather than the main "subject" itself. For example, if you are trying to classify images as either cows or horses, but all of your images of cows are in pastures and all of your pictures of horses are inside barns, your classifier may learn how to distinguish fields from barns (i.e. the context) rather than the difference between cows and horses (i.e. the subjects).

Having well sampled data is a basic requirement, but it can still have other issues. If the data is "dirty" in some way, it may not be useable for actually solving the problem at hand, because it cannot be cleanly processed or there are features that cannot be cleanly tied to the associated events or entities.

Another common issue is when the training data contains information that real-world input data would not have, but is highly correlated with the prediction target. For example, if you were trying to build a model that predicted whether students were going to graduate on time and the data contained information on whether they had completed all requirements, it's very likely that your model will find the correlation between that and a students eventual graduation. In the real world that information would not yet be known and allows the model to "cheat". This phenomenon is called *data leakage* and it can take many forms.

While your training data may be strongly representative of the real world data, it is effectively a snapshot in time of the data distribution. If the events or the environment changes, your model, trained on a past snapshot of the data, will begin to have performance issues. This is a form of data drift. To avoid issues, the performance of the model must be regularly monitored.

Finally, coming back to the first step in the process of building a machine learning model, formulating the problem, it's easy to build a model that solves the wrong problem. This sounds like something that's trivial to avoid, but the reality is that the process of building a model is often complicated and time consuming and the basic problem statement is lost in the confusion of the details. Alternatively, ML modelers are often given a task that is not well defined. In both cases, it's important for the model builders to regularly step back and make sure that the model is providing a solution to the problem that they (and any stakeholders) have set out to solve.

Common machine learning models

There are many kinds of "traditional" machine learning models. Some of the mostly commonly used these days include:

- Linear regression (regression)
- Logistic regression (classification)
- Random forests (regression, classification)
- Gradient boosted tress (regression, classification)
- k-means (clustering)

All of these are made available in popular open source ML software libraries, such as scikit-learn for Python users and others for R users.

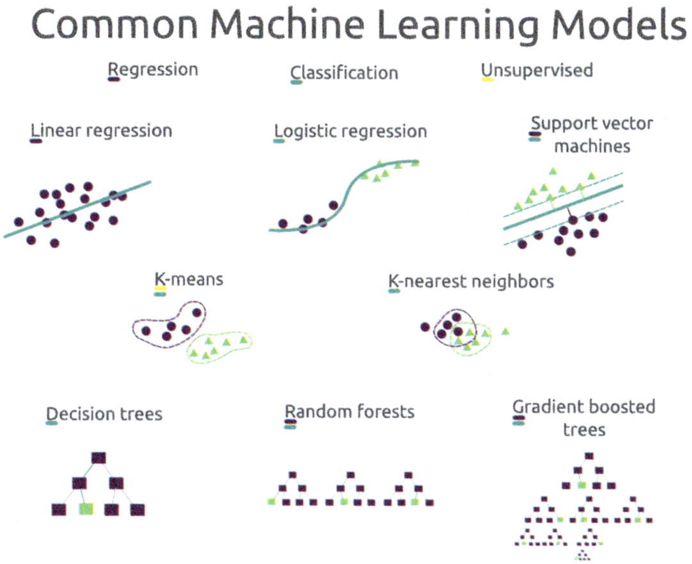

Common machine learning models

From "traditional" ML to deep learning

The remainder of this book is about neural networks and deep learning. Neural networks are are specific type of ML model with a number of interesting attributes, but share most of the basic principals of machine learning found in this chapter. Because of that, I will refer back to many of the concepts presented in this chapter in the subsequent chapters.

References

For your reference, here are a number of resources for learning more about machine learning basics and traditional ML models.

Courses:

- Andrew Ng's introductory machine learning course[14] on Coursera.
- Yaser Abu-Mostafa's machine learning course[15] from Caltech.
- Michael Littman and Charles Isbell's introductory machine learning course[16] from Georgia Tech / Udacity.
- Trevor Hastie and Rob Tibshirani's Statistical Learning MOOC[17] from Stanford.

Books:

- "An Introduction to Statistical Learning: With Applications in R" by Gareth James, Daniela Witten, Trevor Hastie, Robert Tibshirani. Available free as a PDF[18]. Springer Verlag.
- "Data Science from Scratch: First Principles with Python" by Joel Grus. O'Reilly Media.
- "Machine Learning with PyTorch and Scikit-Learn: Develop machine learning and deep learning models with Python" by Sebastian Raschka, Yuxi (Hayden) Liu, and Vahid Mirjalili. Packt Publishing.

[14] https://www.coursera.org/learn/machine-learning

[15] https://work.caltech.edu/telecourse.html

[16] https://www.udacity.com/course/machine-learning--ud262

[17] https://www.dataschool.io/15-hours-of-expert-machine-learning-videos/

[18] https://www.statlearning.com/

Chapter 3

Neural Networks

This book is about "deep learning", but deep learning is really just a name for the modern use of (deep) neural networks. Before we get to deep learning, we need to first look at traditional (shallow) neural networks and understand the basic concepts and issues with them. In this chapter we will cover those basics, preparing us to understand the more recent innovations that fall into the category of deep learning.

What is a neural network?

A neural network, or more properly an *artificial* neural network[1], is a type of machine learning model. Under the surface it's a collection of mathematical operations that take raw input data and, based on the internal structure and parameters of the neural network, it produces an estimate or prediction of some quantity as the output. Just as with other ML models, the parameters of the neural network are learned from training data.

Unlike other machine learning models, neural networks are explicitly inspired by the structure and, to some degree, the function

[1]NB: I will use "neural network" and "NN" to refer to artificial neural networks everywhere else in this book.

of animal brains. They are collections of neuron-like structures that are connected to each other as a network, similar to real neural networks in real brains. The neurons, or nodes, in artificial neural networks are much simpler than real neurons and are designed to have convenient mathematical properties, rather than to simulate or closely mimic the mechanics of real neurons. That said, the basis for exploring these types of models has been the flexibility and power of real brains. For this reason, artificial neural networks became one of the main approaches to the problem of artificial intelligence.

As we will discuss in the rest of this book, neural networks are not one single model, but due to the flexibility of how the components of neural networks can be combined, they form a very large class of machine learning models. We will first look at the simplest form of neural network and in later chapters learn about many of the variations that are used for specific tasks.

What are some tasks that neural networks can accomplish?

Because of the variety of internal structures possible with neural networks, they are able to be applied to many different kinds of tasks. In recent years they have proven to be the best solution to several tasks, including most computer vision and language related tasks.

Some examples of tasks that neural networks perform well on:

- Identifying what is in a photograph
- Recognizing faces and fingerprints
- Transcribing speech
- Translating text from one language to another
- Generating images and text from scratch or from prompts
- Transforming raw images into other styles

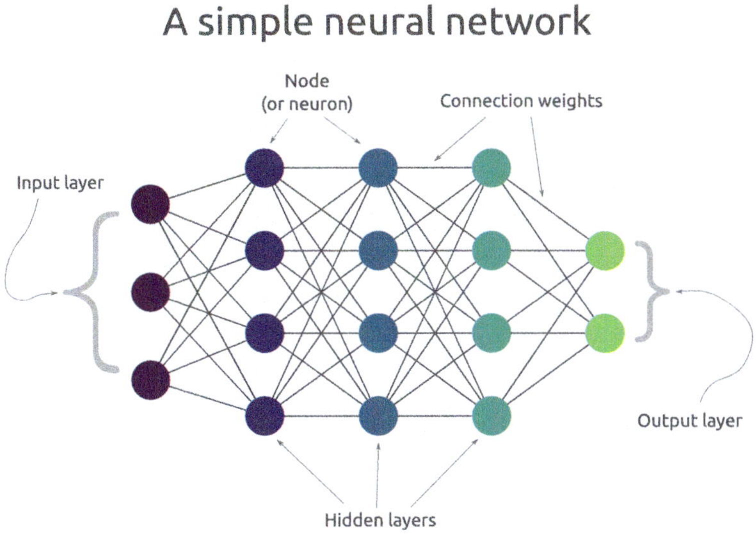

A basic neural network

- Playing games like Atari and go (typically in conjunction with other algorithms)
- Web search
- Product recommendations
- Recognizing and transcribing handwritten or printed text
- Compressing data

Some of these tasks have only become tractable in the past decade due to the development of (useable) deep neural networks. On some tasks neural networks have even been able to match or surpass human performance. Other tasks have been addressed with neural networks for decades. The success and promise of neural networks across so many tasks is one of the main reasons why they are of so much interest to researchers and engineers. They are not one-trick ponies. They can be adapted to almost any machine learning task.

The building blocks of neural networks

Simple neural networks are relatively straight-forward, brain-inspired structures. In fact the simplest version of the neural network is just a model of a single neuron: the perceptron. While not commonly used today, perceptrons laid the groundwork for later neural network structures.

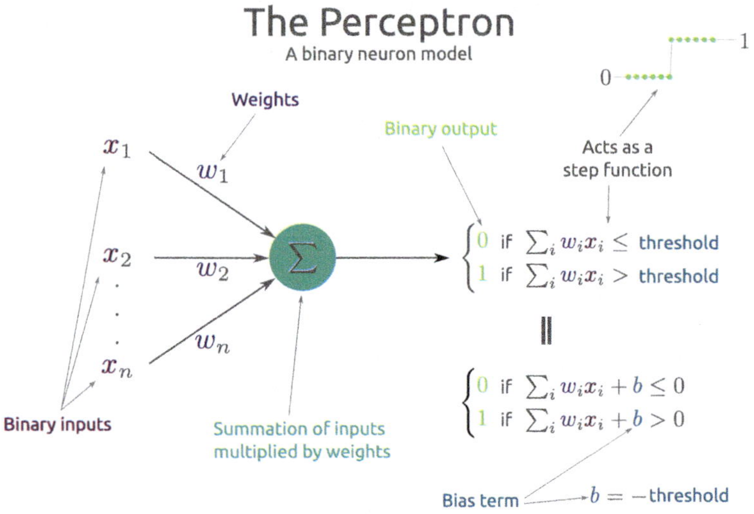

The binary perceptron: a binary neuron model

The original perceptron was designed as a binary digital circuit. It has multiple binary inputs, x_1, x_2, ... , x_n, which can take on values of 0 or 1. These raw inputs are multiplied by corresponding weights, w_1, w_2, ... , w_n. These weighted inputs are then summed up. If the summed value is greater than some threshold, the perceptron outputs a 1, otherwise it outputs a zero. The parameters in the perceptron model are the weights and the threshold. By adjusting these parameters the desired output can be achieved (or at least approached). These parameters can be learned via supervised training, just as with the ML models we looked at in Chapter 2.

$$\text{Perceptron output} = \begin{cases} 0 & \text{if } \sum_i x_i w_i \leq \text{ threshold} \\ 1 & \text{if } \sum_i x_i w_i > \text{ threshold} \end{cases}$$

The perceptron output criterion equation can be rewritten, such that it's the weighted sum plus a "bias" term, equal to the negative value of the threshold. If this total sum is greater than zero, the output is one, otherwise it's zero. This simplified formulation is inline with the notation used for modern neural networks.

$$\text{Perceptron output} = \begin{cases} 0 & \text{if } x \cdot w + b \leq 0 \\ 1 & \text{if } x \cdot w + b > 0 \end{cases}$$

Here $x \cdot w = \sum_i x_i w_i$ and is the inner or "dot" product of the inputs and the weights.

Activation functions

The summing and threshold process in the perceptron is called its *activation function*, which is once again inspired by how real neurons work. In the perceptron this activation function is a *step function*, taking only two values: 0 and 1. For binary classification we only need two values, but even so it turns out that a step function is not the best activation function. A step function has such a sudden change from 0 to 1, that a very small change in inputs or weights can flip the output. In practice smoother functions are more useful for several reasons. One is that relatively smooth[2] functions that are not simply linear can enable the modeling of non-linear patterns. The other reason is related to gradient descent optimization: some functions have derivatives (i.e. slopes) that are more amenable to effective gradient descent optimization.

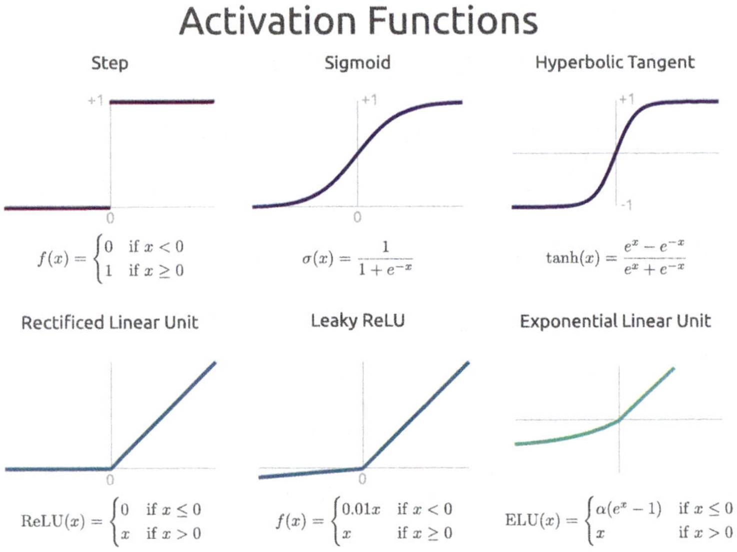

Historical and common activation functions

Historically the most common activation functions were sigmoid functions, which have an S-shaped curve, smoothly transitioning from 0 (or -1) to 1. In more modern neural networks, activation functions such as the *ReLU*[3] are used for reasons alluded to above and which we will discuss in more detail later.

[2]Here I'm using "relatively smooth" to mean functions that are at least piecewise connected.

[3]"Rectified Linear Unit"

Neural network layers

While we have seen what a single artificial neuron, or node, might look like. A true neural network is made up of many nodes connected together, as seen in the first illustration in this chapter. The most common simple neural networks are arranged in layers.

A simple "feed-forward" neural network or "multi-layered perceptron" (MLP) consists of several input nodes, a few so-called "hidden layers" in the middle, and a layer of output nodes at the end. The number of input nodes depends on the input data and how it is encoded. For example, if you had seven numerical features about houses for sale, you could use seven input nodes in the input layer. The number and size of the middle, hidden layers is up to the user to define. And finally, the size of output layer depends on the number of outputs needed. For a classification task with N classes, you would use N output nodes, one corresponding to each class. For a standard regression problem you would only need single output node to provide a single numerical value.

Connections, weights, and biases

Data flows in only one direction in a feed-forward neural network – from the input nodes, through the hidden layers, and to the output nodes. The nodes in each layer are connected to all of the the nodes in the layer immediately before and after it.

The connections represent where the data is flowing, but also the multiplication of that data by weights. As with the perceptron, each connection has its own weight value and each node has its own bias value. The sum of the weight-multiplied data along with the bias term are fed into the activation function. The output of the activation function at each node then becomes the input for the the next layer. These weights and biases are the parameters that the model learns during the training process.

Taken all together, the many weight multiplications, summations, and activation function calls, perform the overall computation of the network, producing the final prediction at the output(s). Data moving all the way through the network once, from input to output, is called a forward pass, and the reason this is called a feed-forward network architecture.

Looking back at the diagram of the simple fully-connected neural network, you can quickly see that as the number of layers grows and the number of nodes in each layer grows that the number of parameters in the network grows extremely rapidly. This is a big part of what gives neural networks their power, but also what can make them difficult to train, as a lot of parameters typically means you need a lot of training data and a lot of computing resources.

Learning via gradient descent

Neural networks learn the best parameters by training on data. Like several other types of machine learning models, they can learn in a supervised manner via gradient descent: iteratively adjusting the model's parameters to decrease the overall prediction error, as measured by the loss function.

> ### A note on math
>
> This section contains some equations, but the important part is really how the equations go together, rather than all of the details of each equation. You should be able to understand the gist without understanding every single part.

As discussed in Chapter 2, gradient descent can be thought of as stepping "downhill" on a "loss surface", which has as many dimension as there are model parameters. The best parameter value combination will produce the lowest prediction error, which corresponds to the so-called "global minimum" in the loss surface. Typically there are many valleys and dips in the loss surface that are lower than the surroundings, but not quite as low as the the global minimum. These are called "local minima". It's not always possible to find the global minium, but there are several strategies to do as well as possible.

To perform gradient descent, the training process needs to know three things: the value of the loss function for a given set of parameters (i.e. at that point on the loss surface), the slope, or gradient, of the loss surface at that point, which indicates which direction to move in, and how big of a step to take.

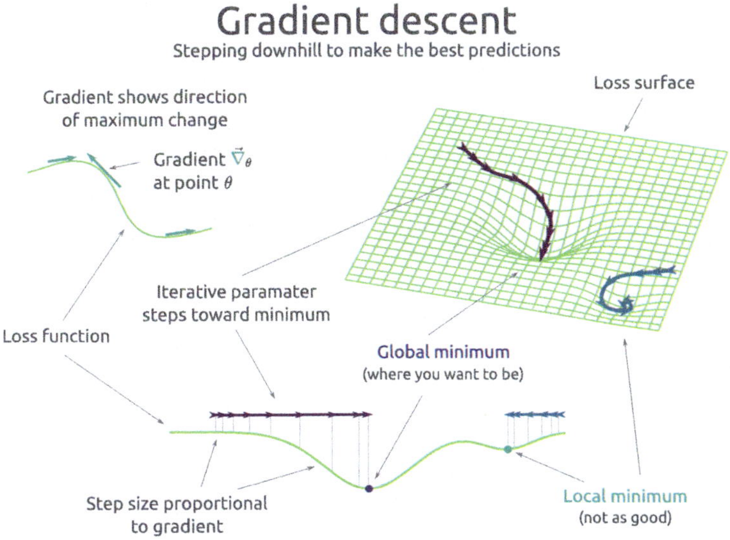

Gradient descent

Evaluating the loss function at a given point is straight-forward: pass the inputs through the network to produce a prediction and compare this with the known "ground truth". Each prediction requires a forward pass through the network. The more data points that are used to make predictions (each requiring its own forward pass), the better the estimate of the loss function, since it is an average value, such as mean squared error (MSE) or mean absolute error (MAE).

The direction of the next step taken is determined by the gradient[4] of the loss surface at that point. The size of the next step is determined by the magnitude of the gradient (i.e. a steeper slope will have a higher magnitude) and the learning rate, which is set by the user.

$$\text{step} = -\alpha\vec{\nabla}_\theta$$

where α is the user-set learning rate and $\vec{\nabla}_\theta$ is the gradient vector at point θ in the parameter space (i.e. for the current set of parameters being evaluated). The learning rate, α, is set by the user and can be a simple number or can be a more complicated rule, such as requiring a smaller learning rate based on the gradient magnitude or on the number of training cycles that have passed.

Backpropagation of the gradient

Determining the gradient is more difficult. The gradient is telling you what would happen to the loss function value if you made a very small change to the parameters in the direction the gradient is pointing in. Mathematically, this is the derivative along each parameter dimension. The gradient is the composite vector, whose components show how the loss function changes when any single parameter is changed.

$$\vec{\nabla}_\theta = \left[\frac{\partial L}{\partial \theta_1}, \frac{\partial L}{\partial \theta_2}, \ldots, \frac{\partial L}{\partial \theta_i}\right]^T$$

[4]Technically you're stepping in the opposite direction of the gradient, as the gradient is a calculation of the vector in the direction of the steepest increase of the loss surface.

The tricky part about determining the gradient is that the loss function depends on parameters throughout the network. A change of a weight in the beginning of the network affects the ultimate prediction output by the network. Fortunately there is a technique that makes the calculation of the gradient relatively easy: backpropagation.

Real world neural networks tend to have many, many parameters, all of which can affect the output. You can think of the output as the result of many nested mathematical operations. The input to the network starting at the innermost level of the nest and the final output coming from the outermost level.

We are interested in the loss function, e.g. $L_\theta = \frac{1}{2}(y - \hat{y})^2$, which is dependent on the set of parameters, θ, indirectly through the prediction produced by the network, $\hat{y}$. To understand how a change in the parameters would affect the loss function, we can therefore look at how $\hat{y}$ is affected by the change in the parameters.

Let's consider the simplest neural network, a network with only one node per layer. This will give us a simpler scenario, while still giving us the core conceptual parts of backpropagation. To determine how a small change of a parameter somewhere in the network would affect the output, we first look at the derivative of the final activation function with respect to its immediate parameters. The final output could be written like

$$f_f(z) = f_f(b_f + w_f \times a_{f-1})$$

where z is just a placeholder for the input to the function, b_f is the bias term of that layer, w_f is the weight of that layer, and a_{f-1} is the output of the activation function of the preceding layer feeding into the final layer. We can keep fleshing this out to see the nested operations and end up with a really long, hard to read equation:

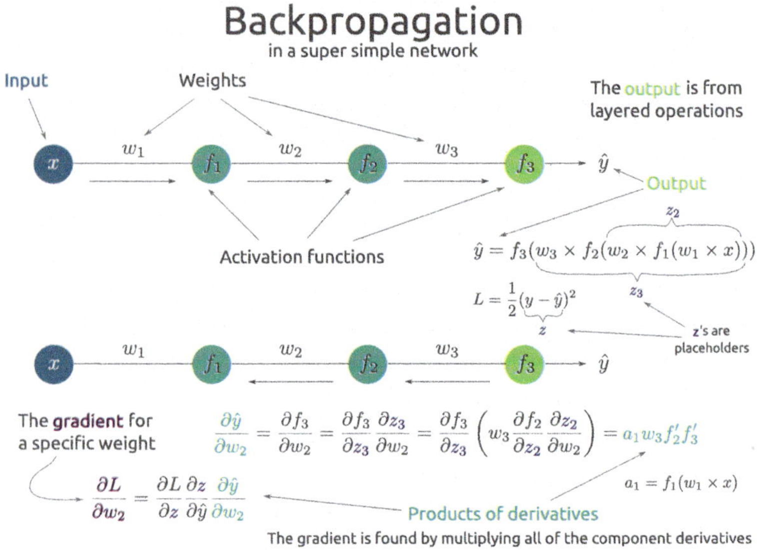

Backpropagation in a simple neural network

$$f_f(x) = f_f(b_f + w_f \times f_{f-1}(b_{f-1} + w_{f-1} \times f_{f-2}(b_{f-2} + w_{f-2} \times f_{f-3}(\ldots))))$$

where x is the initial input to the network and the index indicates which layer the term is from, relative to the final layer. Fortunately, we can understand the concept without having to work through every layer.

The derivative of the the final activation function with respect to its own weight, for example, would be

$$\frac{\partial f_f}{\partial w_f} = \frac{\partial f_f(z)}{\partial z} \times \frac{\partial z}{\partial w_f} = f_f' \times a_{f-1}$$

where z is a placeholder for the input to f_f, i.e. $z = b_f + w_f \times a_{f-1}$. The important thing this equation is telling us that we need to know the derivative of the final activation function with respect to its input and we need to know the derivative of the input with respect to w_f. This works, because the chain rule of calculus tells us that the derivative of a nested or compound function is the product of its component derivatives.

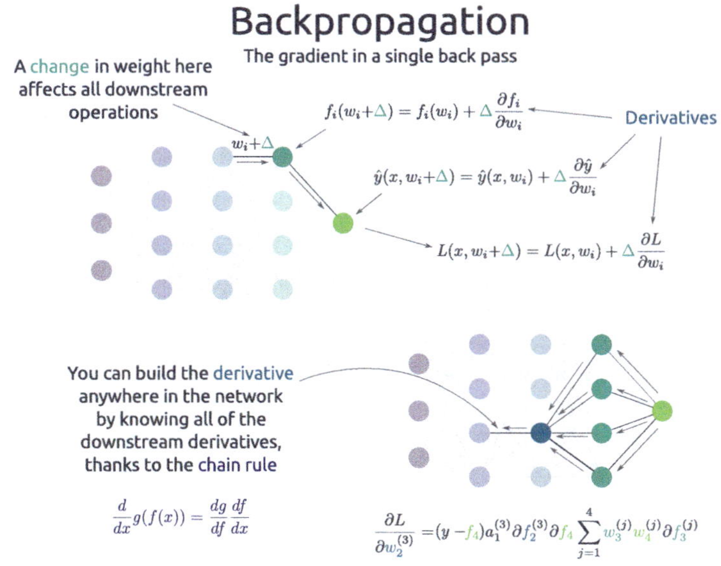

Backpropagation to build the gradient

To find the derivative of some parameter deeper in the network, say $\partial f_f / \partial w_{f-5}$, we need to then find the derivative of each layers that comes after it by working backwards from the final layer, finding the derivative of each components as we go. This is the process of backpropagation. While this example looked at a very simplified neural network, the procedure works with more complicated networks as well.

It turns out that this process is much more computationally efficient than alternatives, such as varying the parameter and estimating the change. Instead we get all of the derivatives throughout the network as part of the backpropagation process in single "backward pass". This was one of the most important early discoveries that made training neural networks feasible.

Why are the loss function and performance metric often different?

A common question is why models are often not trained using the performance metric, such as accuracy, but instead are trained on a loss function such as cross-entropy.

The answer is that for many model types, such as neural networks, it's much easier to work with a loss function that is easily differentiable. Many performance metrics are not differentiable (at least in ways that could be efficiently used for training). You'll instead use a loss function that is easily differentiable, while also serving as a very good proxy for the performance function.

The flip side of this question is: why not just use the loss function as the performance metric? While this is possible, the loss function is often hard to directly interpret and is thus less useful as the metric for many tasks.

Vanishing gradients and parameter initialization

As we just saw, the gradient of a neural network is the product of many terms. One of the problems that can arise from this is the effect of very small or very large gradient values in the early layers, which area the products of many, many terms. In particular, if the gradient has many very small terms, the product will tend to zero, which means that the loss surface is very flat. A flat loss surface means that the network will be very slow to learn, as even big steps will make only small improvements in reducing the loss function value.

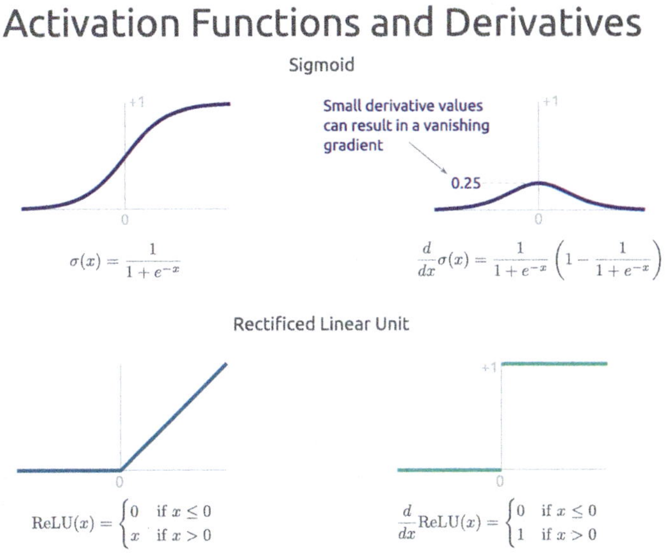

Activation functions and derivatives

Activation functions such as the sigmoid function are particularly susceptible to creating vanishing gradients, as their derivatives max out at 0.25. Any time you multiply lots of positive values less than one, you will be driving the product toward zero. This is one reason why alternative activation functions like the rectified linear unit, or ReLU, have been adopted. The ReLU function has a derivative of one for most values in practice.

Vanishing gradients pose a fundamental problem, as layers that are early in the network will tend to learn much more slowly than layers later in the network and they may not learn at a feasible rate at all.

Another key aspect to learning is initializing the parameters of the network. Even with well behaved activation functions, if you initially set the parameters to extreme values, you can cause gradients to vanish (or explode, which is an equally bad problem, but this is easier to deal with[5]). Therefore, there are many strategies to how one should initialize the parameters of a network. While there are no perfect strategies, the general idea is to set values randomly with a mean of zero and a variance that's smaller as the number of parameters increases, such as in Xavier initialization or He initialization.

Output layers

The final layer of a network is usually different from the middle, hidden layers, as the final layer needs to produce values that are interpretable as solutions to the problem the network is designed to solve. For example, the network maybe predicting the class or label of the object in an image (a classification problem) or it may be predicting the remaining useful life of a tool (a regression problem). For different use cases, different types of activation functions (a.k.a. layers) are used for outputs.

For regression problems a linear output function is appropriate, as it can take on a wide range of values. The output layer would then have as many output nodes as were needed for the prediction, e.g. one node for predicting the sales price of a house versus two nodes for predicting the size of a rectangle that could frame an object in an image.

While it may seem that a step function would be appropriate for a (binary) classification output, a step function does not allow gradient-based learning. Additionally, most classification methods actually produce a value that is more like a probability. So instead of just 0 or 1 in the case of binary classification, a value such as 0.78 is produced and then used either directly or with a threshold

[5]Gradients that are too large can be dealt with via gradient clipping: any gradient magnitude greater than some threshold is set to the threshold value.

for classification. By adjusting the threshold, the user also has the ability to choose how comfortable they are with different amounts of misclassification.

For multi-class classification, such as labeling many different species in images, a function such as softmax is commonly used.

$$\text{Softmax}(x_i) = \frac{e^{x_i}}{\sum_{j=1}^{n} e^{x_j}}$$

Softmax, or soft argmax, gives the probability of the answer being of class i. The sum in the denominator makes the total of all output values for classes 1 through n sum to one, as is required for probabilities. For a multi-class classification network, the final layer would then have as many nodes as there were classes, e.g. 10 nodes for classifying images of single digits 0 through 9.

What does a neural network do?

Having looked at the mechanics of how neural networks work, it's natural to ask "what does a neural network actually do?" in a more abstract sense and "how is a neural network different from other types of machine learning models?".

In most traditional machine learning models, much of the effort from the person training the model is around feature engineering, i.e. creating, iterating, and selecting the best features. The goal is to find the information or ways to represent the information that best allows the model to cleanly separate classes in the data or pick out the pattern needed to predict numerical values (i.e. regression). While some of this can be automated to a certain extent, it is one of the main focuses when training traditional machine learning models.

Neural networks present a different set of capabilities and shift the focus away from feature engineering. One way to think of this is that a neural network is learning how to warp the existing feature "space" in a way that makes the classes easily separable by the final layer of the network. Similarly, when performing a regression task, the network is learning to represent the features in a way that allows it to easily perform a simple linear regression in the last layer. The flexibility within the network allows it to combine the inputted raw features in different ways until it lands on the best way to represent these features for the goal of the network.

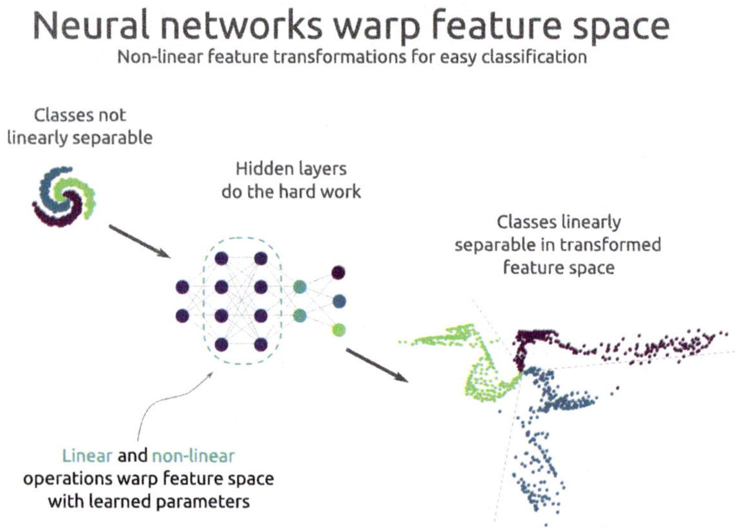

Neural networks transform feature space to make the problem easier to solve

This can be contrasted to traditional methods such as a support vector machine classifier, where the person training the model needs to select a transform kernel that will project the features into a space that allows for easy separation of the classes. Done properly, a neural network will learn the equivalent of this projection kernel on its own.

An important aspect of neural networks that makes this possible is the layer structure of the network. This lends itself to a hierarchical representation of data/features that is natural to many types of data. Thinking about image data, we can imagine the hierarchy of "features", going from individual pixels, to basic lines, to curves, shapes, and complicated structures, such as faces. A neural network tends to naturally learn these types of hierarchical levels in data.

Generalization in a model is about finding the common patterns, while ignoring the "noise" specific to individual examples. One way to do this is by learning "compressed" representations of data. Encoding data in a smaller number of bits typically results in "loss", meaning that it is not quite the same as the original. If you can create a compressed encoding that can then decompress in a way that has relatively high fidelity to the original (by some measure), that typically means that non-essential aspects (i.e. noise) have been removed. Neural networks are often able to create these kinds of compressed representations of data internally, preserving the fundamental patterns and helping them generalize.

As we will discuss later, a final very important aspect of neural networks is their composibility. Because of their layered, hierarchical nature, neural networks can learn to represent data for one task and then be repurposed for other tasks by modifying the later layers in the network. This access to intermediate feature representations from somewhere in the middle of the model is unlike most other types of ML models, where typically only the final result is of any value.

From basic neural networks to deep learning

Traditional neural networks are very flexible and robust, but were not seen as necessarily any more powerful than other traditional machine learning models. The reason for this is probably two-fold:

- There were some difficulties in training larger, more powerful models that had not yet been overcome.
- Large neural networks need a lot of data and computing resources to really shine.

In the remaining chapters of this book we will look at some of the techniques and applications that have led to and come out of the deep learning revolution.

Resources

Some further resources for learning the basics of neural networks and deep learning:

Courses

- The NYU Deep Learning[6] course (Theme 1) by Yann LeCun & Alfredo Canziani.
- Fast.ai's Practical Deep Learning for Coders[7], with Sylvain Gugger and Jeremy Howard.
- Andrew Ng's Neural Networks and Deep Learning[8] on Coursera.

Books

- "Neural Networks and Deep Learning[9]", by Michael Nielsen
- "Machine Learning with PyTorch and Scikit-Learn: Develop machine learning and deep learning models with Python" by Sebastian Raschka, Yuxi (Hayden) Liu, and Vahid Mirjalili. Packt Publishing.

---

[6] `https://atcold.github.io/NYU-DLSP21/`

[7] `https://course.fast.ai/`

[8] `https://www.coursera.org/learn/neural-networks-deep-learning`

[9] `http://neuralnetworksanddeeplearning.com/`

Other online resources

- Neural Networks[10] series by 1Blue3Brown (Grant Sanderson).
- How Neural Networks Work[11] by Brandon Rohrer.

[10] https://www.3blue1brown.com/topics/neural-networks
[11] https://e2eml.school/blog.html#193

Chapter 4

The rise of deep learning

In the previous chapters we have looked at "traditional" machine learning techniques, including neural networks. These methods have proven useful for addressing many predictive problems, but there is a reason this book is about deep learning: deep learning has proven extremely good for certain types of problems that traditional techniques have not excelled at.

Compared to traditional ML models, deep neural networks have proven to be good at a wide range of problems, but have turned out to be particularly good at problems involving computer vision (i.e. image based problems) and natural language processing. In this chapter we will discuss how and why deep learning came to the forefront of machine learning, with a very brief overview of the history of deep learning[1]. In the following chapters we will go into more detail about techniques specifically designed to address computer vision and natural language problems, as well as some generic and advanced deep learning techniques and practical considerations.

[1]As of 2022, the English Wikipedia page on deep learning[2] has a reasonably good overview of the historical events that led to the current deep learning boom. https://en.wikipedia.org/wiki/Deep_learning

Moving to deep neural networks

The most basic definition of "deep learning" is a neural network with more than one hidden layer (i.e. the middle layers of the network). By this definition people have been doing deep learning for a long time, but the deep neural networks achieving impressive results in the past decade have tended to be much "deeper" with tens or even hundreds of layers.

Having more hidden layers has some theoretical advantages over only a few, but primarily researchers have shown empirically that deeper networks enable performance that has not been achievable with shallow networks. From a theoretical point of view, it can be shown that networks with many layers can represent certain mathematical functions[3] with exponentially fewer nodes than would be needed to represent the function with a single hidden layer. From a more empirical point of view, deeper networks seem to have the ability to represent the data in a hierarchical way that fits naturally to many types of data. Additionally with deeper networks, you can create more problem specific network architectures, often combining different layers in modular, purposeful ways, as we will see later.

Researchers came up with many of the key ideas for deep neural networks decades ago, but they were not practical to train or use at that time. Several of the key innovations, such as backpropagation and network architectures well suited for solving computer vision and natural language problems, were created in the 1980s and 1990s. Their potential was not fully realized, however, due to the limited computing resources available at the time (relative to modern computing hardware) and high cost of producing enough training data.

[3]The XOR function being a famous example.

What made deep neural networks possible?

While many of the core ideas for deep neural networks had been around for a while, it was only within the last decade that they really took off. In many ways, deep neural networks were "ahead of their time", as the computing technology of the day was simply too underpowered to demonstrate the potential of neural networks.

The name "deep learning"

Deep learning is synonymous with deep neural networks, but "deep learning" is the more popular term (and incidentally in the title of this book). So why is it called deep learning?

According to Andrew Ng, one of the researchers who popularized deep neural networks, the term deep learning has mostly served as a convenient marketing term. It has allowed for a break from the "old" neural network days, when neural networks often suffered from more hype than they could deliver on, and signifying that we've entered a new era.

Regardless, "deep learning" is the most common name for this set of technologies.

The trends that computing resources have followed, such at Moore's law for transistor density, have been the exponential increase of processor speed, the exponential decrease in the cost of data storage, and the exponential increase in network bandwidth. These trends (and other closely related technology trends) have also led to an exponential increase in the amount of data produced and stored. These trends led to the era of "Big Data", starting in the 2000s, and ultimately enabled the emergence of deep learning as a leading ML technique.

One specific development that helped deep neural networks emerge was the evolution of GPUs[4] from being aimed solely at pro-

[4] Graphics processing units

cessing graphics to enabling more general workflows. Because neural networks can be formulated mathematically primarily as matrix operations, researchers were able to port neural network operations to GPUs, taking advantage of the highly parallel nature of the GPUs. Today GPUs are the workhorse hardware for most deep learning training, while other new processor architectures have been designed specifically to handle neural network processing.

The effect of these trends became apparent to the wider ML and computing communities in 2012, when a neural network won the ImageNet Large Scale Visual Recognition Challenge (ILSVRC) competition by a large margin. Up until 2012, ILSVRC and other similar competitions had been dominated by traditional computer vision feature extraction techniques paired with traditional ML models, such as support vector machines. In 2012 a deep convolutional neural network named AlexNet won the flagship ILSVRC challenge with an error rate more than 10% lower than the second place finishers, a huge performance improvement over the rest of the field and previous competitions.

AlexNet, designed by Alex Krizhevsky (namesake of the network), Ilya Sutskever, and Geoffrey Hinton of the University of Toronto, mostly used components and techniques that had existed previously, but was able to put them together in a way that led to a huge performance increase. AlexNet consisted of convolutional layers, ReLU activations, established dropout for regularization, and was implemented in a way that allowed it to train on multiple GPUs[5]. By combining these established ideas, some newer tricks, and porting the code to run on the latest hardware, they were able to train on a very large dataset and achieve a breakthrough in predictive performance. The dataset itself, ImageNet, was much larger that previously available datasets. The images were sourced from the internet and labeling was performed by crowdsourcing,

[5]We will cover the same components and techniques used in AlexNet in more detail in later chapters.

something previously not available[6].

News of AlexNet's win at ILSVRC in 2012 spread very quickly and by 2014 the challenge was dominated by competitors using deep neural networks. More importantly, not only was the ML community taking notice, but the wider tech world was paying attention and resources started being directed towards further developing and using deep learning methods.

This new interest in neural networks led to many further, rapid developments, quickly improving the state-of-the-art performance on many tasks. Important in this progress was the development of several open source software projects, the open publishing of results (and often the models themselves), and overall the new momentum of progress in the field, which has seemingly grown consistently over the past ten years. Deep learning is now used across a very wide range of academic and industry domains.

Where are we now with deep learning?

Deep learning is now a very established field of both research and practical applications. That said, we are still likely in the early days. There is much that is not understood from a theoretical point of view regarding *why* deep learning works and what its limits might be. In areas such as natural language processing, researchers have been building ever larger models, reaching the scale requiring supercomputer level infrastructure. It's yet to be seen if the approach of ever-larger models and ever-larger training datasets will continue to yield reasonable improvements. On the other hand, much research is also around how to make models (relatively) smaller and more efficient in training and inference.

In several areas, such as computer vision and natural language processing, deep learning provides not only the best predictive performance, but has become established as "best practice", with easy

[6] `https://image-net.org`, Li Fei-Fei, et al.

to use tools available. While more fundamental research is ongoing, there is also huge effort now around making deep learning solutions practical and scalable for a larger and larger set of users in the real world. What might have previously been an R&D effort requiring skilled researchers to experiment and develop has become closer to simply importing a few robust open source libraries by typical software engineers.

One of the big changes from the world of "traditional" ML models is that deep learning moves the user away from feature engineering to "network engineering", i.e. finding the best network configuration for the problem at hand. At this point the research and user communities have landed on a relatively established set of network architectures to use many common tasks. Importantly, for many common tasks there are available pre-trained models that can be reused or adapted via transfer learning (more on this later).

The last ten years have been a remarkable period of progress for neural networks (and machine learning in general) on all fronts: there are far more people working on it, far more resources, and far more real-world use cases than ever before. No one (or neural network...) can predict the future, but it seems that we are still in the early days of this technology.

Chapter 5

Computer vision and convolutional neural networks

Computer vision is one of the most important application areas of deep learning. In fact, it was advances in computer vision that kickstarted the current era of deep learning. Further advances and applications of deep learning to computer vision tasks have dominated this first decade of deep learning[1]. Recent deep learning techniques have led both the cutting edge and best practices for real-world applications in almost all areas of computer vision.

In this chapter we will look at computer vision tasks and the network architectures and techniques that apply to them.

Computers and images

Working with image data has been an important task of computers for decades. The explosion of digital photography has made this area even more important, as images are now overwhelming born in digital form and the shear number of images has grown enormously in the past two decades.

[1] As we will see, techniques for other types of tasks, such as natural language processing, have more recently made huge strides in performance and have gained considerable interest on par with computer vision.

Historically, computer vision has referred to techniques that required a high-level understanding of the contents of an image. This was seen as distinct (or perhaps a subset) from digital image processing and graphics programming. The current application of neural networks to image-related tasks has somewhat blurred these distinctions. In this book we will use "computer vision" to mean any image-related task that neural networks can be applied to, from image classification to image generation and beyond.

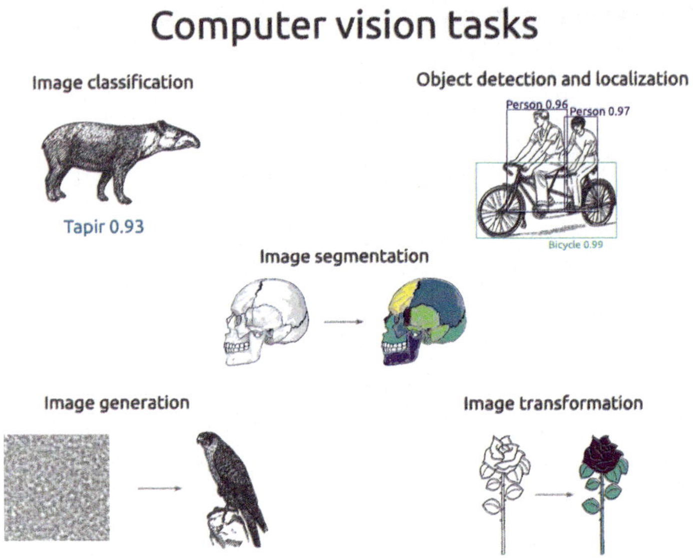

Common computer vision tasks

Computer vision tasks

Before we discuss deep learning techniques for computer vision, we will first look at several common categories of computer vision tasks. Understanding what these tasks are and how solutions are evaluated is important for understanding how deep learning is applied to those tasks. First we should note that digital images are simply arrays of numerical values, where the numerical values represent the data from each pixel. For typical color photographic images, an image will have multiple two dimensional arrays of data, one for each of the colors in the format – usually red, green, and blue layers.

Image classification is the task of identifying what the main subject of an image is. Some examples include labeling a photograph as containing a dog versus a cat, detecting whether a medical image contains a tumor, determining whether an image from a traffic camera shows congestion, or determining the species of a plant in a photograph. These can all be formulated as standard binary, multi-class, or multi-label classification tasks. There are several metrics for assessing model performance on classification tasks, such as accuracy, precision, recall, F1-score, Matthew's correlation coefficient, log loss, and Area under the ROC curve. Each has tradeoffs and care should be taken in choosing the best one for the task at hand.

Object detection and localization is a task related to image classification. Instead of merely predicting whether an object of some class is in the image, object detection and localization involves finding where in the image an object of one of the known classes is. Examples include a phone camera detecting faces to adjust the focus, detecting vehicles, signs, and pedestrians in images from a self-driving car, detecting tumor tissue in a biopsy slide image, and tracking balls in video images from sports. The target for the prediction is typically the bounding box[2] around the object and the class label of that object. This means that the model needs to do two things at once and thus needs to be evaluated for both goals simul-

[2]Rectangular bounding boxes are the most common prediction target for localization tasks, but other polygons or paths denoted by a list of points can also be the target.

taneously. Average precision is one of the most common metrics to measure detection and localization.

Image segmentation is the labeling of every pixel in an image with some class. Examples include delineating what is and is not the road in a traffic image, outlining the different organs in a medical image, classifying each pixel as foreground or background in an image, or labeling different types of ground cover in satellite images. Labeling each pixel as belonging to a class, such as identifying which pixels are grass, trees, and sky, is called "semantic segmentation". Another type of segmentation is "instance segmentation", where the goal is to label each pixel as belonging to specific instances of an object, such as counting the oranges on a tree. A special case of segmentation is depth estimation, where instead of classifying each pixel with a label, each pixel is given an estimate of distance from the camera, making it a pixel-wise regression problem.

For semantic segmentation, the most common metrics are intersection over union (IoU) and the Dice coefficient for binary segmentation tasks and mean IoU and mean Dice coefficient for multiclass tasks.

Image transformation is the task of changing an image in some way. This is a broader category that has overlap with several other categories. Examples include colorizing a black and white image, "super resolution", where an image's resolution is increased by "intelligently" filling in new pixels, fixing problems in images, such removing an object from an image, or filling in an area of an image, changing an image from one "style" to another, and adding elements to images, such as whimsical effects and filters to selfies. Tasks such as colorization are very similar to segmentation, mentioned above, as the task is to provide a label (i.e. color) or numerical value (i.e. pixel-wise regression) for each pixel. Many image transformation tasks involve several steps, such as detecting an object and then modifying the image in some desired way, such as automatically blurring faces in images. Evaluation of image transformations is very task specific.

Image generation is the creation of entirely new images or filling in regions of images in some desired way. Examples include generating landscape scenes, with or without input from the user, altering the pose or facial expression of a person in an image, or generating models wearing clothing products. Image generation and related tasks, such as image transformation, require metrics that measure how similar the generated image is to being a real image. We will look at some of these later when we look at generative adversarial networks.

Traditional computer vision

Solutions to computer vision tasks have been researched and developed for decades. Most traditional approaches rely on techniques developed for digital image processing tasks, such as filter techniques to find edges in an image. In general the most typical approaches prior to the deep learning era incorporated manually optimized feature extraction techniques combined with traditional machine learning models, such as support vector machines.

Many of the traditional feature extraction methods, such as SIFT, are very powerful and fast to run, but they are typically highly specific in application and brittle, not handling different data gracefully. The big innovation in neural network-based approaches was the ability of the model to learn how to extract features on its own directly from the training data. Instead of the user trying to extract specific features, the user typically only cares about the final, overall performance on the task and lets the model figure out what features to extract on its own.

What's hard about computer vision tasks?

Compared to humans, computers have historically been very bad at computer vision tasks. It's extremely difficult to hand-code the logic to tell a computer how to do most of these tasks. Describing the difference in appearance between a cat and a dog using words (or computer code) is almost impossible, but of course our eyes and brains can easily do this, as they evolved to carry out visual tasks very well.

The geometric, yet irregular nature of images make them difficult to deal with. Most settings for real world images have relatively low constraints, i.e. there can be great variation of how the same subject appears in an image depending on numerous conditions, such as lighting, distance to the lens, orientation of the subject, etc. This is why for tasks such as face recognition, it's advantageous to require the image setting to be as uniform as possible, such as those required for ID photos.

There are some more fundamental difficulties in computer vision tasks. For example, you may be able to hard code a detector that finds triangles in an image, but the logic may fail if the triangle is shifted, rotated, or skewed somehow. As humans we know that the triangle is still a triangle despite being moved to another part of the image, but capturing the essence of "triangleness" may be very difficult in code. This is one reason for the desire to create models that can learn how to determine "triangleness" on their own by looking at lots of examples of triangles (or really objects that are much more complicated than simple geometric shapes). If the fundamental patterns of these objects and their common variations can be learned, a much more robust model can be built.

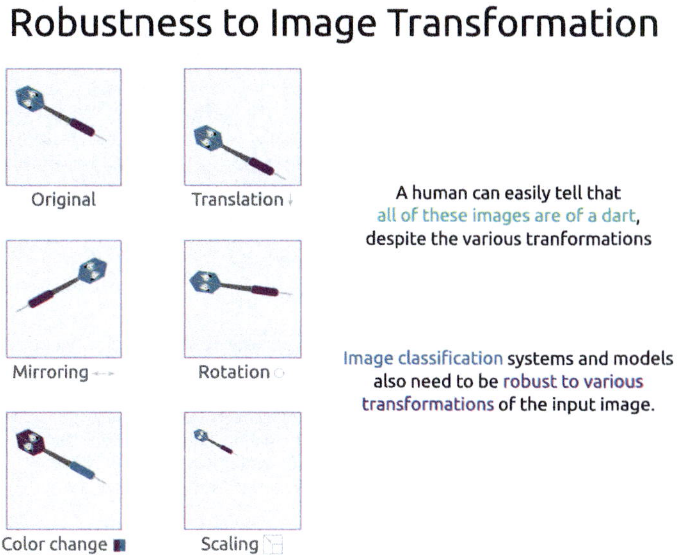

Robustness to image transformations

Convolutional neural networks

Multi-layered convolutional neural networks (CNNs) presented a solution to handling these image-based tasks. As we'll see, convolutional layers can learn to extract features on their own, can detect these features regardless of location in the image, and when stacked together can take advantage of the hierarchical nature of most image data.

Convolutions

Convolutions are (unsurprisingly) the core operation of convolutional neural networks. A convolution is a mathematical procedure to apply a single operation over all parts of an array of numbers. As

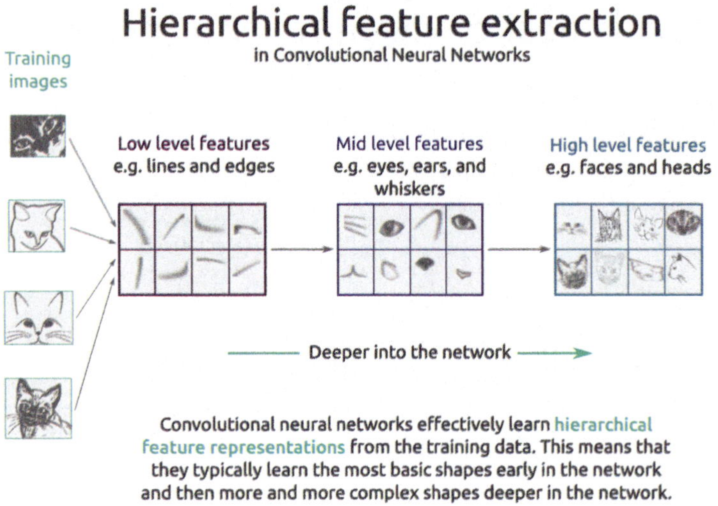

Hierarchical feature extraction in CNNs

mentioned previously, digital images are simply arrays of numerical data. The convolution process can be thought of as multiplying another (typically smaller) array of values, known as a "filter" or "kernel", with a part of the target array (for us an image). This multiplication is done element by element. The product of this multiplication is then saved and the filter is moved to another part of the image, saving that result, and further repeating until the filter has been applied to all regions of the image. The output array is also known as a "feature map".

Mathematically, a convolution is relatively straight forward. For a one dimensional filter applied to a one dimensional array, the convolution[3] can be represented as

[3]This formula is technically a cross-correlation, rather than a convolution, but

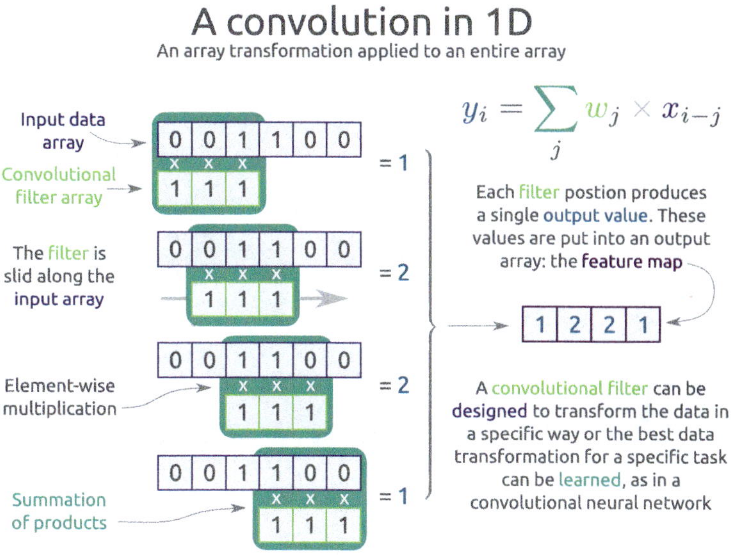

A convolution in 1D

$$y_i = \sum_j w_j x_{i+j}$$

where w_j is the filter weight for position j in the filter array, x_{i+j} is the (image) array value at $i + j$, and y_i is the value of the convolution operation from position i of the (image) array. As i is increased, it's as if the filter is sliding along the (image) array and you get the result of applying the filter to the entire array.

A two dimensional filter applied to a two dimensional (image) array is conceptually similar, but the filter is moved both across and down the (image) array.

---

it produces an equivalent output and is the form used in practice most of the time.

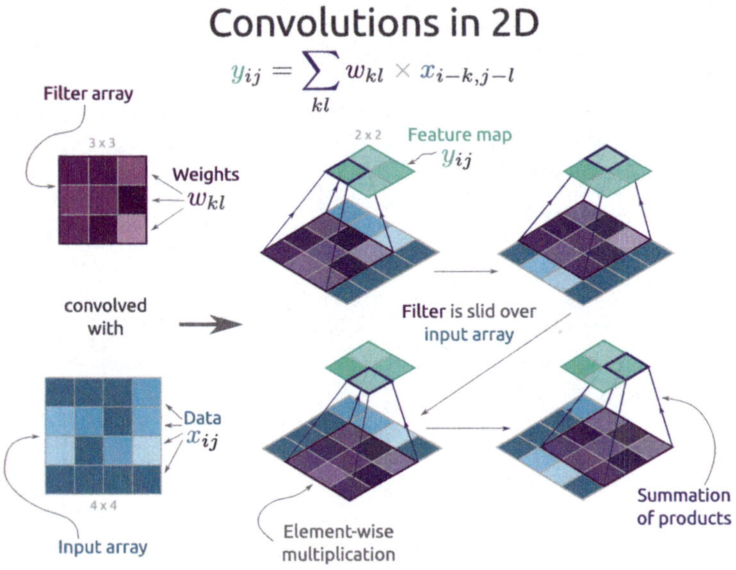

A convolution in 2D

$$y_{ij} = \sum_{kl} w_{kl} x_{i+k,j+l}$$

The most important part to understand is that the same filter array of weights is being multiplied with each different window region of the (image) array. The question then becomes "what are the best weight values in the filter array?".

Traditionally convolutional filters have been a workhorse of digital image processing and were designed by hand to do things like edge detection or blurring. The beauty of neural networks is that the weights in these filter arrays can simply be learned from the training data. While traditional filter arrays have been carefully designed to perform a specific task, we do not ultimately care what an individual filter in our neural network does, only that it helps the network output the best prediction. This hands-off approach was a big break from traditional image processing.

Filter size, strides, padding, and pooling

There are several knobs that can be turned to tune convolutional filters in a CNN. These are typically treated as hyperparameters of the network.

Filter size is the dimension of the filter array – how large the window is that you're sliding over the image. A smaller filter will focus on smaller features, whereas a larger filter will focus on larger scale features.

A **number of filters** can be applied to an input array simultaneously in parallel. This is somewhat analogous to the number of nodes in a standard, fully-connected neural network layer. Each filter is applied independently, learning its own weights, and producing its own feature map.

Stride is how far the filter is moved each time you slide it along. A "vanilla" filter would be moved one pixel over, having a stride length of one. A stride length of three would move the filter over by three pixels each time. Since the number of outputs from performing a convolution depends on how many positions the filter is applied to, a larger stride will produce fewer outputs, and thus reduce the resolution of the resulting array from the convolution. Selecting stride length depends on the desired effect and/or computing resource constraints or efficiency goals and can be treated as a hyperparameter.

Network size, parameter sharing, and computation

One of the tricky parts about dealing with image data is that images tend to be big, i.e. need a lot of memory and storage. Neural

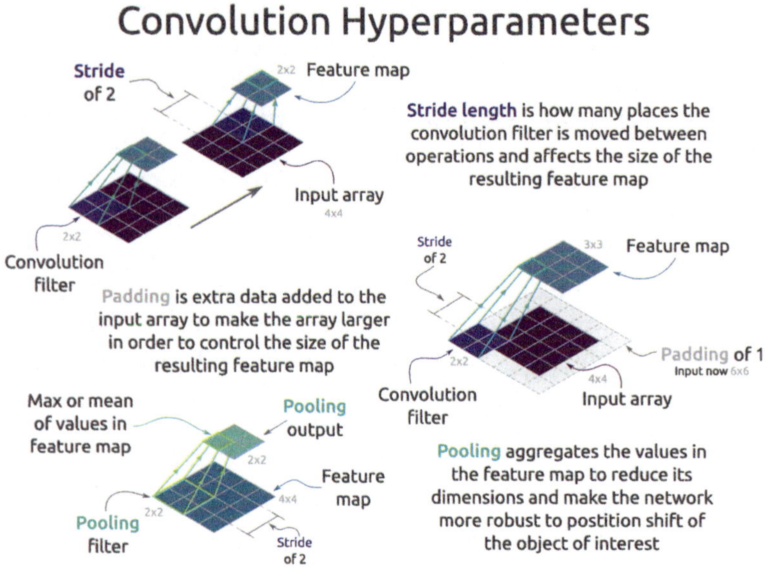

Some hyperparameters of convolutions

networks can have the same issue. The more parameters, the more computing resources, especially memory, are needed to train and run the network. One of the advantages of task specific networks, such as CNNs, is that they can be designed to reduce the number of parameters in the network in clever ways.

CNNs reuse the same filter weights across the entire image, rather than learning separate weights for each part of the image. This is largely to be able to detect specific features (a.k.a. motifs) that can be found in any part of the image, but the side effect is to reduce the overall size of the network.

Of course the general trend in deep neural networks is that bigger is better: more layers allow you to find even more robust patterns. But, you want those layers and the deep network to be as resource efficient as possible.

Padding means adding extra data (pixels) to the edges of your image. Unless your filter is of size 1x1 and stride 1, the result of the convolution will be a matrix of smaller size (i.e. lower resolution) than the original image. Additionally, the pixels at the edges of an image are seen less by the convolution operation compared with pixels farther in. Padding can alleviate these issues. The amount of padding depends on the goal, e.g. maintaining the same resolution, which is termed "same padding". The most common data value to use for padding is zero.

Pooling is the reduction of the size of the output array (a.k.a downsampling) by performing an operation, such as taking the mean or the max, on the data values in subarrays of the output. The values are "pooled" together. This in turn reduces the resolution of the feature map as well as the number of weights needed downstream. Conceptually, pooling is a way to compress the information contained in a feature map, maintaining the larger scale features while discarding some of the "noise". Pooling also adds some robustness to where in an image specific features are found, i.e. an object not in the center will be more like to be detected as the same object as if it were located in the center of the image. A pooling filter of 2x2 pixels with a stride of 2 is a common size.

A basic CNN architecture

The convolution layer is the most important building block of convolutional neural networks, but a single convolution layer isn't able to do everything on its own. A typical CNN consists of multiple convolution layers, plus several helper layers. In this section we will look at a basic CNN architecture and what the different components do as well as the effect of combining them all together.

A single convolution layer learns to detect (or extract) some important features from an image. We don't know at the outset what those features will be, as the network tries to learn the most useful features to extract for the task it's given. Empirically, it's been observed that stacks of CNN layers tend to extract increasingly complex features, starting with basic shapes, such as straight edges, and then simple shapes, and finally complex features, such as faces or vehicles. Each subsequent layer builds on what features have already been extracted.

A very basic classification CNN might have three convolutional "blocks", a basic unit of convolutional operations, and two fully connected layers at the output to handle the classification.

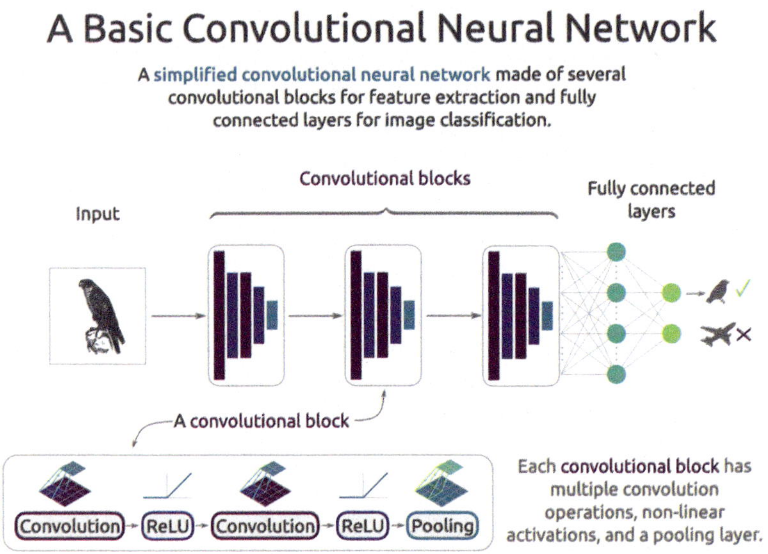

A basic convolutional neural network

Each convolutional block might have several convolutional layers, with an activation function, such as a ReLU, in between the convolutions, and finally a pooling layer. Typically the dimension of the feature maps decreases with each convolution. By stacking several convolutional blocks together, the network can learn more and more complex features to input into the fully connected layers at the end of the network.

There are many architectural and hyperparameter choices that can be made to modify this most basic architecture. Researchers and engineers have empirically found architectures and hyperparameter choices that have worked well for certain tasks. Some of these choices are better understood than others from a theoretical point of view.

Some important CNN model architectures for computer vision tasks

Historically there have been several computer vision networks that have demonstrated and popularized certain architectures for specific tasks. These are important touch points for understand the evolution of computer vision networks and how different architectures relate to each other.

AlexNet

In the "modern era" of deep neural networks, AlexNet[4] was the one that is considered the most important from a historical point of view. It is not a network that is still used, but its win in the 2012 ILSVRC Top 5 classification competition was the event that kicked off much of the current deep learning era.

AlexNet is similar to the basic CNN described earlier, with six convolutional blocks, using ReLU and max pooling, and two hidden fully connected layers at the end, followed by a softmax layer for making multi-class classifications.

ResNet

One of the next milestone CNN architectures was ResNet, a classification CNN designed by Kaiming He, Xiangyu Zhang, Shaoqing

[4]See the discussion of AlexNet in Chapter 4 for more historical context.

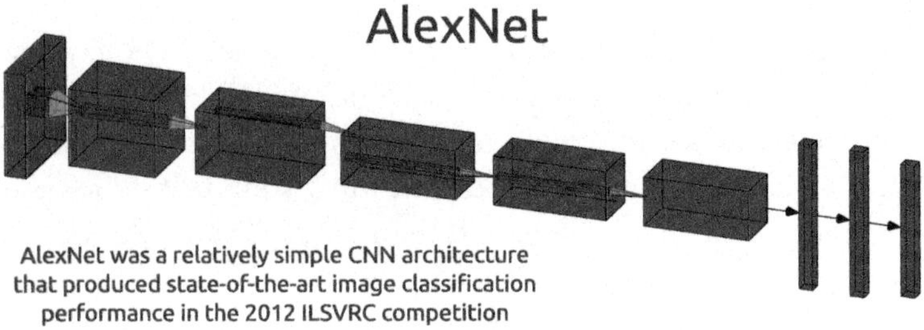

The AlexNet architecture

Ren, and Jian Sun for the ILSVRC 2015, which it won.

The ResNet architecture, an abbreviation of "Residual Network", is based on residual connections, which let some of the data bypass the convolution layers, such that it can be used unaltered by downstream layers. Researchers had noticed that after some number of layers, adding more layers actually made error on the *training* data (and test data) worse. He, et al, proposed that theoretically this should not happen, as any additional layers could just learn to pass the data through unaltered (i.e. learn the identity function[5]). In practice, though, this was not happening. He, et al, hypothesized that instead of learning the identity function, it would be easier for the network to learn the difference between the input data (for a layer) and the optimal output data: the residual. To achieve this they added so-called "skip connections", allowing data to bypass some number of convolutional layers. That input data is then summed with the output data of the convolutional layers.

While He, et al, didn't invent skip connections, ResNet went on to win the ILSVRC 2015 competition and popularized the idea

[5]The identity function is simply a function that returns the input unaltered. In matrix math, the identity matrix is all zeros, except the diagonal elements, which are all ones. Learning this very specific matrix has proven difficult for network layers.

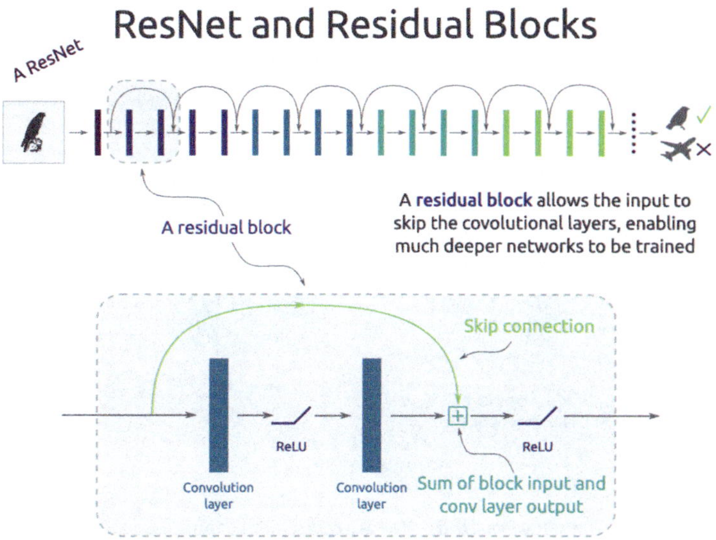

ResNet and residual blocks

of skip connections and residual blocks. Architectures based on these ideas are still the most commonly used CNNs, used for image classification and as the "backbone" of networks designed for other vision-related tasks. Some of the successor architectures to ResNet include DenseNet, ResNeXt, and ResNeSt.

Autoencoders

An autoencoder is a network designed to learn how to encode or compress an image (or other input). For images, it's essentially two CNNs, with the second flipped around. The first "encodes" the input image by learning features and squeezing the resolution down to some smaller size. The second part (a.k.a. the "decoder") rebuilds

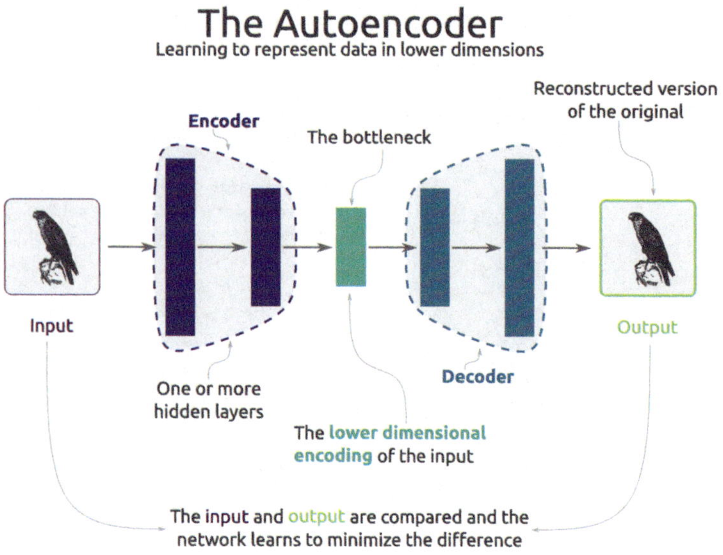

The autoencoder

the image back to its original resolution and contents using an up-sampling method[a]. The difficulty in doing this is that you're trying to represent the full image with a lower amount of data (i.e. compression).

An autoencoder can be trained with unlabeled data, where the input and output are compared (an example of "self-supervised" learning). The network attempts to reconstruct the input with the lowest error possible.

Historically this was a method to "pre-train" layers within CNNs on unlabeled datasets, which were typically larger than the labeled dataset. The convolutional layer from the encoder could then be re-used in the CNN, where further supervised training would take place. Larger labeled datasets, more powerful computing resources, and more efficient CNN architectures made this approach less common, but similar approaches have become very important in other areas, such as natural language processing.

See the section on "transfer learning" later in the chapter.

[a]See "transposed convolutions" below.

U-Net for semantic segmentation

Semantic segmentation is another task on which convolutional neural networks have be able to greatly outperform their non-deep learning predecessors. In a typical semantic segmentation task, every pixel in the input image needs to be labeled as one class or another. One of the most important architectures that has emerged for segmentation is the U-Net. U-Net was introduced in 2015 by Olaf Ronneberger, Philipp Fischer, and Thomas Brox for use in biomedical image segmentation, but it has since been adopted for segmentation tasks across many domains.

The U-Net architecture has three main, important components:

1. The "downward" encoding arm
2. The "upward" decoding arm
3. Skip connections from the encoding arm to the decoding arm

Image data is fed through several convolutional layers in the encoding arm, reducing the size of the image data. After the image / feature data reaches the bottom of the encoding arm, it is sent through transposed convolutions to increase the resolution at each set of layers in the decoding arm. Information from earlier in the network is passed to the decoding layer blocks from the encoding arm to provide more "guidance", in a somewhat similar manner as the skip connections seen in the ResNet architecture. The skip connection data is concatenated to the downstream data from the decoding arm, rather than simply summed, in contrast with the ResNet architecture. The diagramatic way that the encoding

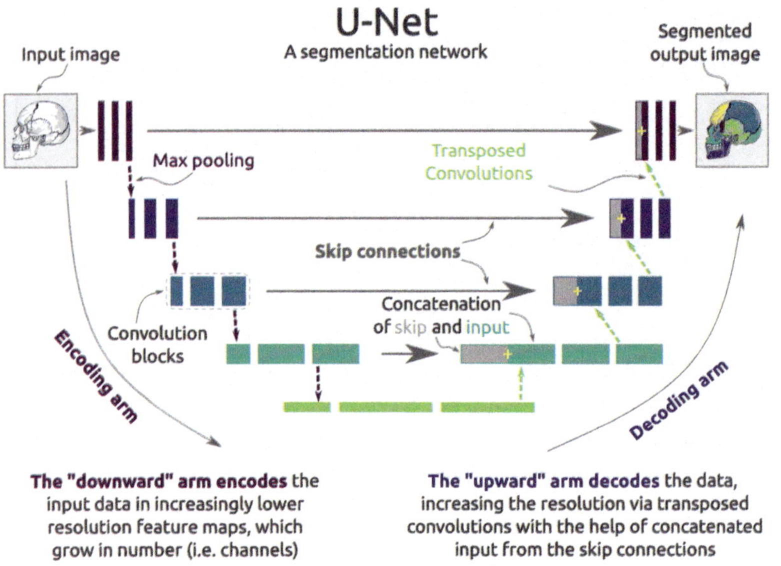

U-Net

and decoding arms are represented is where the name U-Net comes from, as the diagram looks similar to a "U".

While the original U-Net used vanilla convolution in the convolutional blocks, the "backbone" of the network can be made of residual blocks or other similar architectural elements. There are several variations of U-Net which have been created to improve performance or work better for specific segmentation or other tasks. Additionally, the U-Net architecture is used as a component for several other networks.

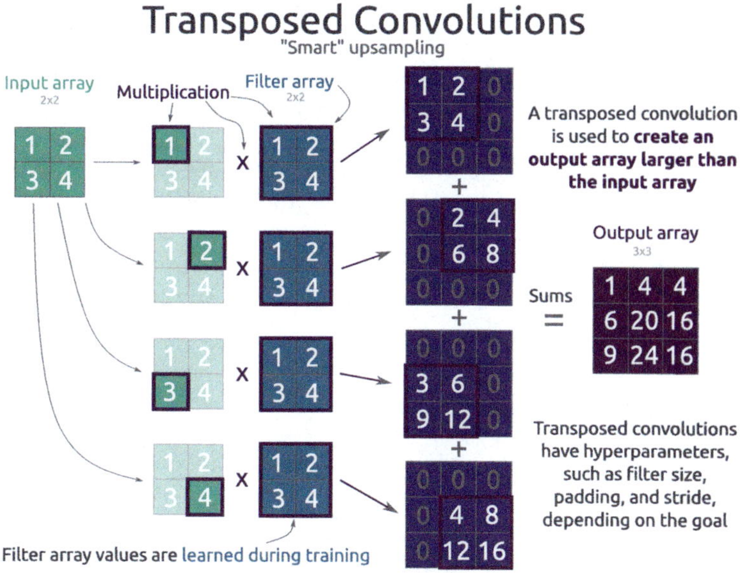

Transposed Convolution

Transposed convolutions

Unless specifically designed not to, most convolution operations result in a feature map that is smaller (i.e. lower resolution) than the input (image) data. Several tasks require the opposite result, i.e. increasing the resolution. One way to increase the array size of the data is with a "transposed convolution".

A transposed convolution applies a filter to the input data in such a way as to produce a larger output array. It has similarities to the convolution process, but also differences. The filter array is applied to only a single cell of the input array at a time, such that the output is the value of the input cell multiplied by the value of each cell of the filter. These values are then added to the output array in a specific position. The filter array is then moved to the next cell of the input array and the process is repeated, with these values added to the output array in the corresponding position of the input cell. The values put into the output array that overlap are simply summed.

Like a normal convolution the values of the filter array are learned in the training process, such that they produce the best values to achieve the overall task of the network. As with convolutions, there are a few key hyperparameters that can be chosen to achieve your desired goal, including filter array size, padding, and stride length.

YOLO for object detection

Object detection involves detecting whether an object from any class of interest is in an image, predicting what class the object is, and estimating the coordinates of its bounding box (or other shape). YOLO is an object detection model that has proven to be one of the best performing object detection models in recent years.

As with image classification, there has been a quick evolution of object detection methods. The most basic approach is to consider many subregions of an image, for example by sliding a window over the image, and run a classifier on each subregion to predict the presence of an object of the class of interest. In principle this works, but it comes at a high computational cost.

YOLO (You Only Look Once) is an efficient object detection algorithm that uses several tricks to make it both accurate and light on resources. It was introduced by Joseph Redmon, et al in 2015. YOLO breaks up the image into a grid and uses convolutional blocks to detect objects everywhere in the image in a single pass (hence, "you only look once"). Training data must be labelled in a YOLO-specific way to reflect which grid cell it belongs to, the presence or absence of an object, the object class, and the boundary box coordinates. Grid cells without objects still need labelled data, though the details of the class and bounding box don't matter.

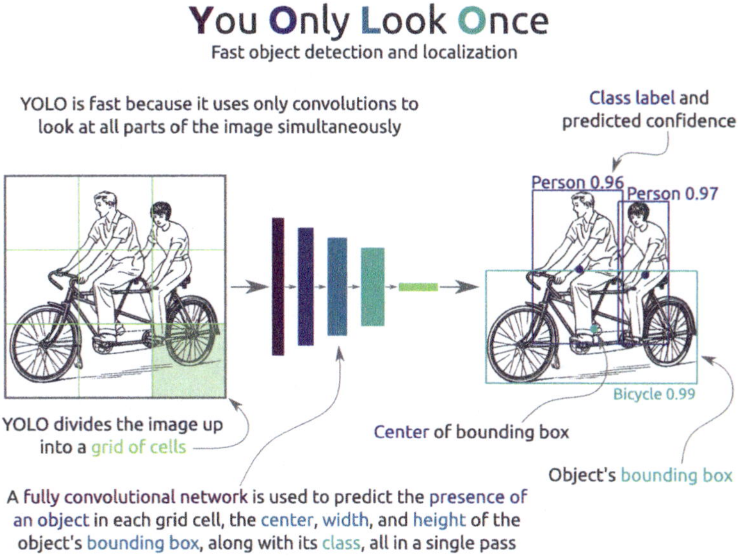

You Only Look Once (YOLO)

The model is set up to predict the center point of bounding boxes, as an object may extend into multiple grid cells. When setting up the YOLO model, the user can choose the number or cells and the number of object centers to look for in each cell, as there could be more than one object centered in a cell. This is a limitation and trade off of the model, but is largely dictated by the type of data. Very busy images with a lot of objects close to one another require more grid cells and / or more object centers per cell.

An object detection model, such as YOLO, may end up predicting several bounding boxes for a single object. To decide which bounding box to choose, a complimentary algorithm called *non-maximum suppression*, or NMS, is used. NMS looks at both the confidence of the model's prediction for a certain class and the overlap of candidate bounding boxes, as measured by intersection over union (IoU).

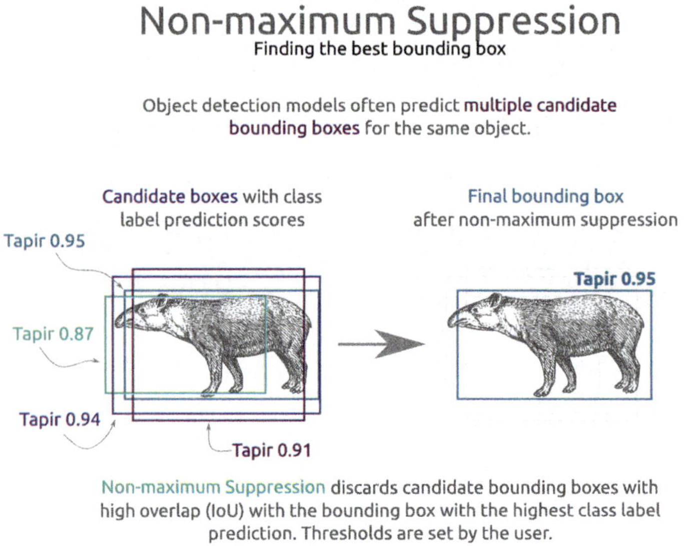

Non-maximum suppression

NMS consists of the following steps:

1. Only objects predicted with a confidence about some threshold are considered.
2. For each class, select the bounding box with the highest prediction score.
3. Calculate the IoU[6] of all other bounding boxes of that class against the previously selected bounding box.
4. Discard all boxes above a user-defined IoU threshold (e.g. 0.5). The boxes with a high IoU score likely bound the same object.

[6]See aside about how IoU is calculated.

5. Of the remaining boxes for the current class, take the one with the highest prediction score and repeat steps 2-4 to find other instances of the same object class.

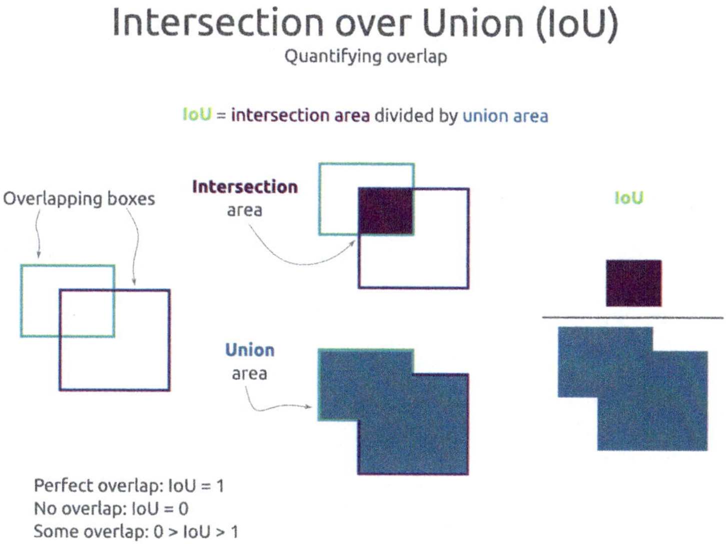

Intersection over Union (IoU)

Intersection over union (IoU)

IoU is a way to measure the amount of overlap of two regions. It comes up in a few different contexts in computer vision and is important to understand.

IoU measures the amount of overlap of two objects, divided by the total area that the two objects take up together (i.e. the *intersection* of the objects divided by the *union* of the objects). Perfectly aligned objects will have an intersection that equals the union and thus an IoU of 1. Objects that don't overlap will have an intersection of zero and thus an IoU of zero. Partially overlapping objects will have an IoU somewhere in between.

YOLO is one of the most important object detection architectures. There are multiple versions of YOLO (some not directly related to the first version) and several other networks that descend from YOLO or incorporate its ideas.

Image generation with GANs

Image generation is the task of creating new images from scratch, modifying existing images, or filling in areas (a.k.a. inpainting) of existing images. In 2014 Ian Goodfellow et al introduced the generative adversarial network (GAN), which became the basis for a large family of neural network architectures that could generate images extremely well. GANs have primarily been applied to images, but have also proven useful for generating some other types of data, such as audio.

GANs consist of two main parts: the "generator" and the "discriminator". Each is a separate neural network. The generator is a network which produces an output image, while the discriminator is a network that tries to classify images as either coming from the training set or having been produced by the generator. Training the two networks becomes a game, with the generator trying to learn how to produce images that can fool the discriminator into thinking that its images are actually from the training set, while the discriminator is trying to learn how to detect the false images. This is the "adversarial" part of GANs. The game between the generator and the discriminator is what makes them able to learn so well, but it can also lead to *instabilities* in the training process.

Improvements to the basic GAN framework have yielded very high quality images that are often difficult for humans to recognize as being synthetic. As we will discuss in Chapter 7, GANs and other

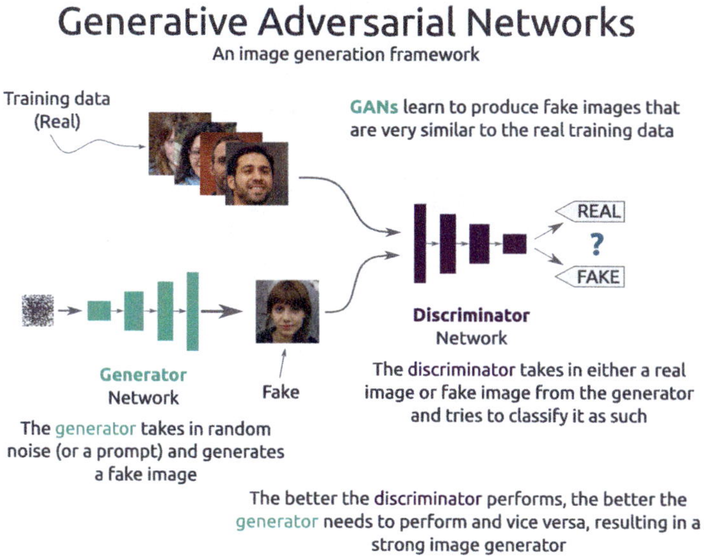

Generative Adversarial Networks

image generation models can be designed to use user input to guide image generation toward a desired output. Other notable image generation architectures include variational autoencoders (VAEs), flow-based models, and diffusion models, that latter getting a lot of attention so far in 2022. We will look more at diffusion models in Chapter 7.

Transformers for computer vision

In the chapter on natural language processing (NLP) we will devote part of it to discussing transformers. Transformers were developed to handle NLP tasks, but have been adapted to also handle computer vision tasks, and have been gaining traction, in part, because they

An image of a person generated by StyleGAN

allow you to use a single architecture for a very broad range of tasks.

I won't go into any detail about how those work right now, beyond saying that most of the transformer based vision models treat images as sequences of image chunks. Transformers for computer vision are definitely worth paying attention to.

Common CNN techniques

We've discussed some of the most important and common network architectures for computer vision. There are also several common techniques that models use that are worth being familiar with, because they pop up so often. Here I will touch on some of the most important ones.

Regularization

The goal of all machine learning models is to learn the general patterns of the real-world distribution of data from the (limited) sample of training data. The opposite of that is overfitting. As discussed in Chapter 2, regularization is one of the approaches for reducing the chances of overfitting to the training data.

There are several flavors of regularization. Two of the most commonly used for neural networks are *dropout* and adding constraints to parameters within the loss function (e.g. L1 and L2 regularization).

Dropout

Dropout is a relatively simple approach to making a network more robust to variations in the data. Nodes within the network are randomly set to zero during training with a probability, p, a hyperparameter of the model. By "turning off" these nodes randomly, the network must learn to perform its task despite losing some of its connections. Since zeroing out nodes reduces the overall weight in the network, non-zeroed values are multiplied by $1/p$ to scale the parameters back up to the level that would exist without the dropout.

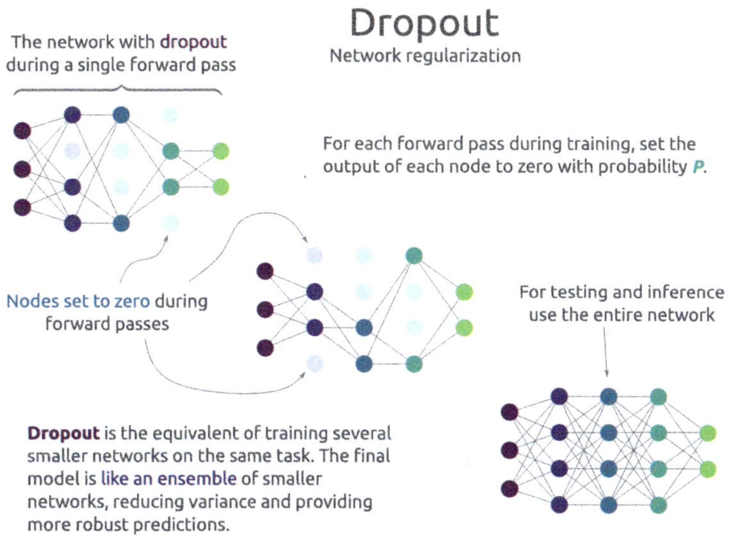

Dropout

Zeroing out nodes is effectively the same as creating several smaller, independent neural networks, the outputs being combined in an ensemble, which is a general technique for reducing variance.

L1 and L2 Regularization

As discussed in Chapter 2, L1 and L2 constraints are commonly used to regularize certain types of models. They can also be used with neural networks, often being referred to as *weight decay*. L1 and L2 regularization both work by adding an additional term to the loss function, where the sum of the model's weights are multiplied by a hyperparameter, often denoted by λ. For L1, the absolute value of the weights is used. For L2, the square of the weights is used.

By placing a term incorporating the total magnitude of the weights in the loss function, the network is forced to "budget" the weight values on only the most important nodes and connections. This means that it cannot as easily fit noise within the training data, as it doesn't have an unlimited weight budget. See Chapter 2 for more detail and an illustration of how this type of regularization affects overfitting and generalization.

Data augmentation

Models are typically able to better learn the general patterns and avoid overfitting if they are trained on more data. More data is almost always better. Since data can often be "expensive" to obtain, however, one approach to increasing the size of the training set is by synthetically creating new, labeled data. For images, this can be as easy as flipping the image right to left. For most scenarios, that type of change would result in the same label on the data.

There are many ways in which image data can be augmented: moving, rotating, stretching, scaling, cropping, blurring, or flipping the image, adding noise, changing the colors, etc. The important part of augmenting the training set with data transformations is

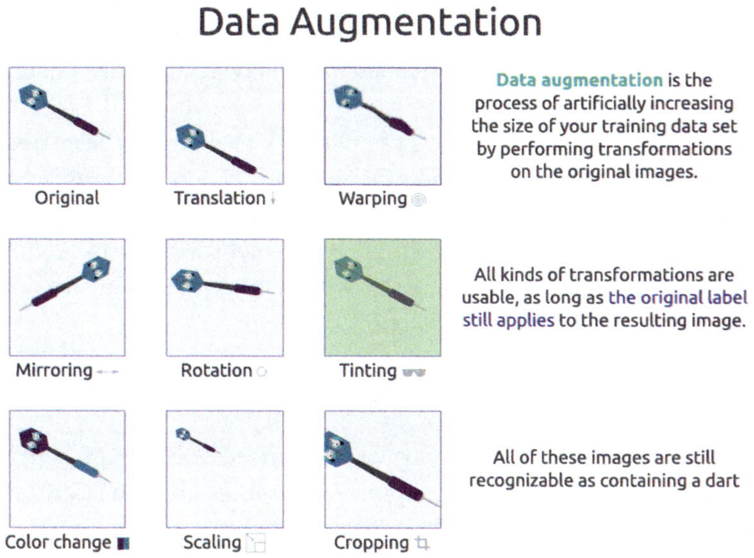

Data Augmentation

that the new image is not changed so much that its label is no longer appropriate.

By automating this process, a training data set may be increased many-fold.

Batch normalization

Training tends to work better in neural networks if the different input features are on similar scales, such that similar step sizes mean the same thing during gradient descent. This is accomplished by "normalizing" the input features, assuring that the different features all have means around zero and a variance around one.

It turns out that if you also normalize the data between each layer, it helps to improve the robustness of the model and speed up training. In 2015 Sergey Ioffe and Christian Szegedy proposed a technique called "batch normalization" to normalize the data before each new layer (or wherever the *batchnorm* layers are placed in the network).

Batch normalization allows the network to learn the best "normalization"[7] for the data inside the network. Each batch normalization layer learns the best parameters for this normalization. Empirically it has proven useful, though the exact mechanism of why it helps is debated. It's often used with convolutional layers and fully connected layers.

Gradient descent algorithms

Gradient descent is at the heart of supervised training of neural networks. As discussed in Chapters 2 and 5, gradient descent allows neural networks to find the best parameters for the network based on the training data by taking steps "down hill" on the multidimensional loss function surface. The down hill direction is found by calculating the gradient for each feature dimension and the parameter values are updated via the process of backpropagation.

It turns out that "vanilla" gradient descent has some limitations in practice, primarily the speed of the process and the amount of resources needed to handle large data sets (and large data examples, like images). In order to overcome some of these limitations, several variants of gradient descent have been developed. Here we will look at a couple of the most important ones.

Stochastic gradient descent

Stochastic gradient descent (SGD) is a form of gradient descent that uses a single example from the training data to perform gradient

[7] Batch normalization is not strictly a normalization, but rather a transformation of the data, starting with a normalization and then followed by scaling and shifting using learned parameters that seem to give the best transformation.

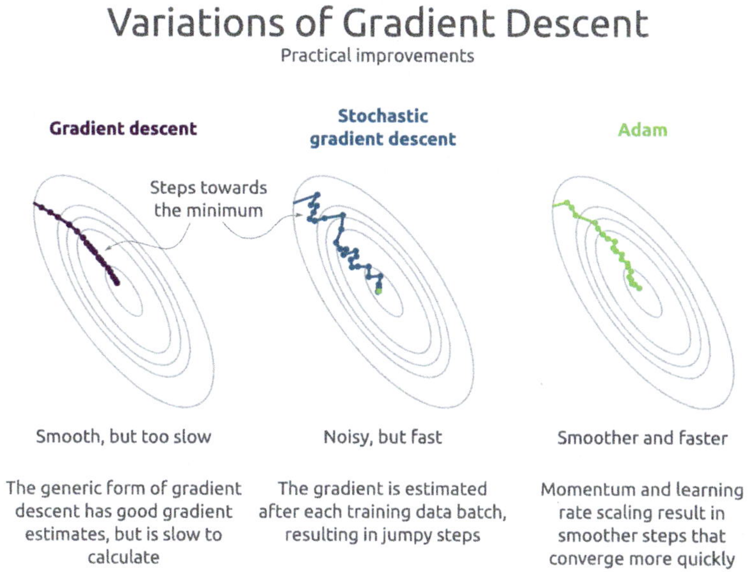

Variations of Gradient Descent

descent. Instead of calculating the value of the loss function by running the model on all examples in the training set, SGD instead calculates the value of the loss function on a single example. Inherently, this means that the estimate of the loss function is worse than averaging the loss value from all available data, but it means that it can be performed very quickly. This "quick and dirty" estimate of the loss function means that the subsequent update (or step) in the parameter values will likely be sub-optimal – not necessarily the best direction or step size. This is the "stochastic" part of SGD.

It turns out, that by taking many steps based on the single data example estimate of the loss function and thus the gradient, the model's parameters will *still* end up moving in the right direction. It's like a random walk, but with the wind blowing the walker in the right direction. Training may need more steps, but the overall training time can be greatly reduced. This has proven to be a very useful approach in practice.

While SGD was formulated using one example at a time, it's more commonly done with a small number of examples or "batch" of data, sampled from the training data set.

Adam

SGD is not without its own limitations. The direction of the next step in SGD is based purely on the gradient of the current mini batch. Due to the small size of the batch, that can mean the next step is in a very different direction than the previous step. While this will work out over time, researchers have realized that if the direction of the next step is less abrupt, the training is likely to converge to the best parameter values more quickly.

One mechanism for taking smoother steps is called "momentum". With momentum the process takes steps as if the gradient was more like a force acting on a moving mass, which wants to maintain its current trajectory. This is accomplished by essentially taking a moving average of the previous steps along with the current gradient estimate. The magnitude of the momentum is a hyperparameter.

Another issue that SGD has is related to learning rates. If the same learning rate is used the entire time, SGD is prone to overshooting good minima. Intuitively, gradient descent is a bit like golf. In the beginning you (typically) need to take long shots and then as you approach the hole, you take short shots. This can be better achieved for SGD by dividing the learning rate by a term related to the size of recent steps. If the most recent steps have been large, the learning rate is scaled down. Techniques called AdaGrad[8] and RMSProp[9] both used this kind of scaling on a per parameter basis.

[8] Duchi, John; Hazan, Elad; Singer, Yoram (2011). "Adaptive subgradient methods for online learning and stochastic optimization". JMLR. 12: 2121–2159.

[9] Hinton, Geoffrey (circa 2011). "Lecture 6e rmsprop: Divide the gradient by a

Adam is a popular gradient descent technique which incorporates both momentum to smooth out the steps and per parameter learning rate scaling to speed up optimization. Besides learning rate, Adam has two hyperparameters related to these modifications to SGD (as well as batch size).

Transfer learning

One of the most important techniques used in deep learning is so-called "transfer learning". Transfer learning is the process of re-purposing an already trained model for a new task.

Imagine if you had a CNN that had been trained to tell the difference between images of cats and dogs, but you wanted to distinguish images of different breeds of cat from one another. If you had enough well labeled data and computing resources, you could simply train another CNN on your cat dataset, setting up the last layer to predict how many breeds of cat you had. Alternatively, you could try to reuse the existing model by modifying only the final layer and training that layers on a small sample of your data.

This process is called transfer learning. Because the existing model has already learned to extract features and transform the input data in ways that are useful for the original task, it's likely that the features and transformations will be useful for related tasks. The closer the second task is to the original, the more likely that transfer learning will be successful.

On a practical level, transfer learning has been very important because it has allowed users with lower levels of resources to create models that meet their needs. Many researchers and others who have built models requiring large amounts of data and computing resources have made these trained models available for free to the community. Many of the deep learning toolkits and libraries now include pre-trained models out of the box.

---

running average of its recent magnitude". p. 26.

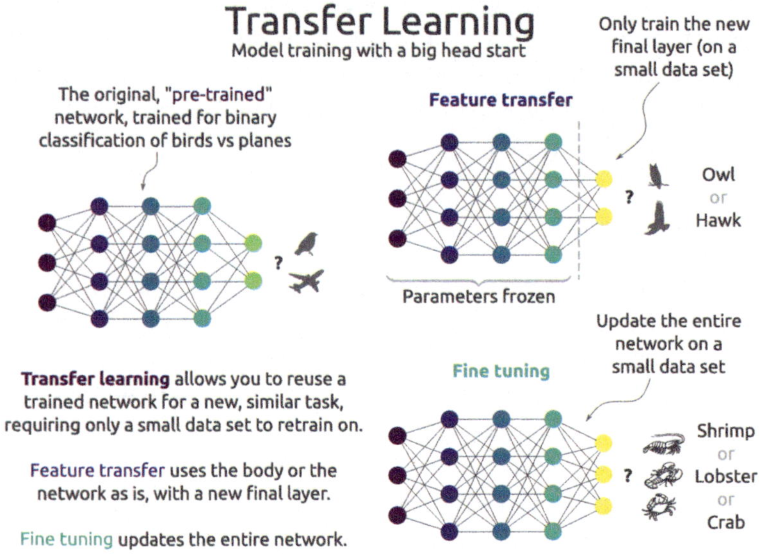

Transfer Learning

Feature transfer

There are a few ways to "re-use" models trained for one task on another task. The most straightforward method is called "feature transfer". In this case the majority of the network is reused as is, but the final layer is modified to fit the new task (e.g. new output classes or moving from a classification task to a regression task). This new final layer is then trained on a small(er) set of data, utilizing the features produced by the rest of the network, which it had previously learned for its original task. In this scenario, the parameters of the rest of the network are said to be "frozen" while the final layer is trained.

Fine tuning

While feature transfer is relatively straightforward, it may not always produce a model with the level of predictive performance needed. A second approach is called "fine tuning". With fine tuning the original model's parameter values act as a starting point for further training of the model. This can be thought of as the pre-trained model giving the final model a big head start in training. Instead of the parameters randomly starting somewhere very far away in the optimization space, they're starting (hopefully) somewhere very close to where they need to be.

For fine tuning a pre-trained model, lower learning rates are often used, as you don't want to overshoot the best parameters by starting with steps that are too large. Like all training, it is as much an art as a science.

Summary and resources

Computer vision has arguably been the most successful application of deep learning in the past decade. Success in computer vision due to advances in convolutional neural networks kicked off the current deep learning wave and the huge increase in research and development of these techniques has led to similar advances in other areas of application of deep learning.

In this chapter we touched on some of the most important concepts, techniques, and models related to computer vision and deep learning for computer vision. There is of course much that was not covered and many details were left out of the topics that were covered, but this chapter should have given you a good jumping off point for further learning, understanding, and applying these techniques.

Below are some resources to help you improve you understanding of computer vision and deep learning techniques for computer vision even more.

Courses

- Fast.ai's Practical Deep Learning for Coders[10], with Sylvain Gugger and Jeremy Howard.
- Andrew Ng's course on Convolutional Neural Networks[11] on Coursera.
- The NYU Deep Learning[12] course by Yann LeCun & Alfredo Canziani.

Books

- "Zefs Guide to Computer Vision[13]", by Roy Keyes. Coming 2023.
- "Neural Networks and Deep Learning[14]", by Michael Nielsen
- "Machine Learning with PyTorch and Scikit-Learn: Develop machine learning and deep learning models with Python" by Sebastian Raschka, Yuxi (Hayden) Liu, and Vahid Mirjalili. Packt Publishing.

Other Online Resources

- How Neural Networks Work[15] by Brandon Rohrer.

[10] https://course.fast.ai/

[11] https://www.coursera.org/learn/convolutional-neural-networks

[12] https://atcold.github.io/NYU-DLSP21/

[13] https://zefsguides.com

[14] http://neuralnetworksanddeeplearning.com/

[15] https://e2eml.school/blog.html#193

Chapter 6

Natural language processing and sequential data techniques

Natural language processing (NLP) is another area where deep learning has emerged as the state of the art approach in the past decade. In this chapter we will look at some of the approaches to handling NLP tasks and other sequential data tasks. Specifically, we will look at recurrent neural network architectures (RNNs) as well as attention-based architectures (transformers).

Text, natural language, and sequential data

Language is one of the defining features of humans. Our ability to communicate not only basic, common things, but also complex and novel topics is one of the characteristics that sets us apart from other intelligent species. As such, the desire to harness computers for handling tasks related to language has existed from the early days of computers, with specific tasks, such as the famous Turning test[1] and language translation, being worked on in the 1950s.

[1]The Turing test was a thought experiment devised by Alan Turing to test whether a computer system had intelligence equivalent to a human. In the test, an external viewer is charged with observing a text conversation between a computer and a human. If the observer cannot reliably distinguish which is the computer

There are many tasks related to language and computers that are considered valuable or interesting. Among those tasks are:

- **Spelling correction**, where known spellings are offered up as alternatives to unrecognized words.
- **Grammar correction**, where "correct" grammatical rewording is offered to the person writing.
- **Sentiment classification**, such as trying to classify whether a product review is positive or negative about the product.
- **Style improvement**, where alternative versions of a phrase or sentence are offered to the person writing.
- "**Named entity recognition**", where known people, places, events, or other important *nouns* are detected in text.
- **Part-of-speech tagging**, where the words and phrases in a sentence are classified by the grammatical role they play.
- **Text and speech user interfaces**, where a human can use natural language to work with a computer, such as a chat bot.
- **Language identification**
- **Language translation**
- **Text generation**, such as summarizing an article or offering reply suggestions to an email.
- **Word suggestion**, such as mobile phone text input systems.
- **Speech transcription**
- **Speech synthesis**
- **Document classification**, such as whether a message is spam.

Some of these are relatively simple tasks for computers, such as spelling correction, while others, such as text generation and translation are harder problems to solve, as they require much stronger understanding of the patterns of language.

More general than natural language tasks are sequential data tasks. Sequential data is any data that has an order, where the or-

---

and which is the human, then the computer is considered to have passed the test.

der provides part of the information contained in the data. Besides (most) language, this includes things like:

- **Audio signals**
- **Sensor data**, such as from medical, vehicle, and equipment monitors.
- **Video**
- **Computer logs**, such as website traffic.

Some of the tasks that people are interested in related to non-language sequence data include:

- **Mode change and anomaly detection**, such as whether a patient's heart rate has changed in a significant way.
- **Audio synthesis**, such as generating music in a desired style.
- **Object tracking and prediction**, such as trying to predict where another vehicle will go on the road.

All of the above have the commonality that the data and its order provide important contextual information for the task. This could be granular, such as the letters in a word, or larger scale, such as themes in an essay. All of this provides important context for accomplishing the task of interest.

Types of sequential tasks

The above tasks can be described in very generic terms by looking at the size and role of the sequence in the task.

You can break down these tasks into the following categories:

- **Many-to-one**, such as sentiment classification
- **One-to-many**, such as image captioning
- **Many-to-many**, such as translation

Depending on the category of sequence problem, different approaches are needed.

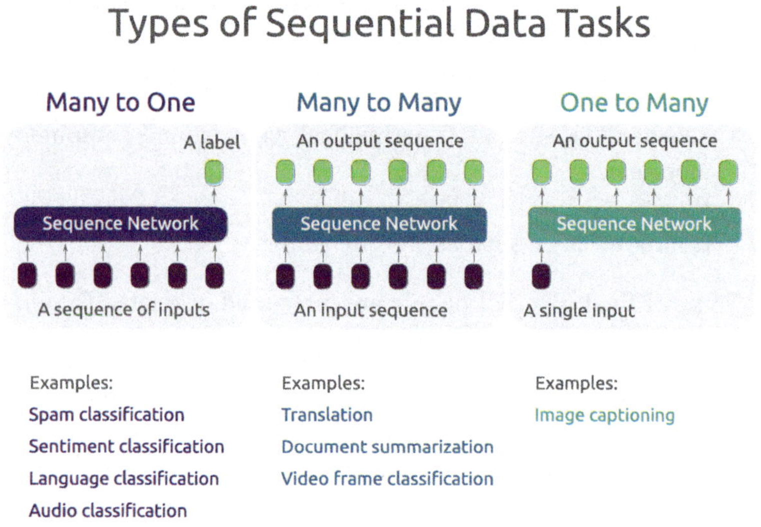

Types of sequential data tasks

Traditional approaches

Compared to many modeling tasks, natural language tends to be "messy". The inherent flexibility of natural language means that a single word can have many variations, sometimes retaining the same meaning and other times modifying the meaning. For example, "dog" and "dogs" are simple singular and plural variations of the same word, but "dog" can also be a verb or used as an adjective, "dogged", which are related, but distinct from the canine animal noun sense of the word. You can easily imagine how complicated things can get.

While languages generally follow grammatical rules, there are numerous exceptions to those rules, making it extremely difficult to capture all of the edge cases that exist in real world usage. These kinds of flexibility make many natural language tasks very difficult for computers.

Traditionally, NLP and sequence-based tasks have relied on methods using some combination of hand-coded logic and statistical understandings of language data. Raw language and sequence data typically needs to have a lot of preprocessing performed to make it usable by computers and to create features that can be used for making predictions or decisions.

Preprocessing and explicit feature creation can be as basic as splitting text into words based on spaces or as complex as calculating word frequencies and distributions. Much of the work related to using traditional NLP methods is around creating these usable features and removing low information words, such as very common words like "the" or pause words like "uh" in speech.

Some commonly seen traditional techniques for text related tasks include tokenization, stemming, lemmatization, identifying stop words, creating n-grams, calculating term frequency inverse document frequency (TF-IDF), edit distances, and bag of words and word count. Some of these are also used with deep learning techniques and we will touch on those later. For other types of sequence data, autoregressive features are commonly created, such as sliding means.

With hand-engineered features, traditional ML methods, such as support vector machines and tree-based models, as well as non-ML models, such as hidden Markov models and Bayesian techniques, can be applied to NLP problems. Those same techniques can be applied to other sequential data problems. Time series related tasks are often approached with so-called "time series methods", such as ARIMA, which typically do not have learnable parameters.

Making a neural network remember

Natural language and sequential data tasks have long been tackled with non-deep learning techniques, but it turns out that there are deep learning techniques that are even better suited to many NLP and sequence tasks.

The hard parts about natural language, in particular, and sequential tasks are the very large number of possible sequences (as either inputs or outputs) and the need to understand "context". By context, we mean the important parts of the data surrounding any specific piece of data under consideration.

If you read the phrase "I saw it", it's unknown what "it" refers to without context. In this case the context is any preceding text that allows you to understand what "it" is referring to. This means that a computer needs to be able to understand and remember that context in order to perform any task that would require taking "it" into account.

Neural networks, as we have seen them so far, are "stateless". They don't retain any kind of memory of the inputs they've seen previously[2]. So how do we enable a neural network to remember the context it has already seen?

The recurrent neural network

A recurrent neural network, or RNN, is a type of network that is able to process a sequence and capture the context to make its predictions. In order to do this, the RNN needs to be able to step through the sequence while retaining "memory" of the items in the sequence it has already seen.

One way you could achieve this is by making several copies of a neural network and having each one process a different step in the sequence, while also looking at the output of the network that processed the proceeding item in the sequence.

---

[2]You could argue that ML models have some memory of data they have seen during training, but the models we have seen so far have no way to retain memory of inference-time, non-training input data that has previously been passed through them.

The raw output of the network that has processed the previous item in the sequence plays the role of memory and is typically referred to as the *hidden state* of the network. This would allow this chain of networks to process the sequence in order, extract features from the inputs, and maintain some memory to establish context.

Why not use a standard feed-forward network?

Neural networks are very flexible and in theory can tackle almost any modeling task you throw at them. As we've seen with computer vision, though, task specific architectures tend to be more efficient and ultimately higher performance than generic feed-forward networks.

There are a couple of issues with using standard networks for sequence tasks:

- The length of the inputs and outputs are not always fixed for the same task.
- A lot of parameters would be needed, so you lose out on parameter sharing that can take place in a specialized sequence architecture like an RNN.

RNNs and other sequence architectures are designed to deal with these issues efficiently.

While this might work for some sequence tasks, where the inputs and outputs have fixed lengths, there is a much better way to do this with only a small change to this "chain of networks" architecture: adding recurrence.

RNNs are essentially chains of networks, but instead of having several sub-networks, they use the same network for each network in the chain, feeding the raw output of the network back to itself in order to process the next step in the sequence. This is the recurrence. This means that each step in the sequence shares the same parameters and that the network can handle sequences of variable length.

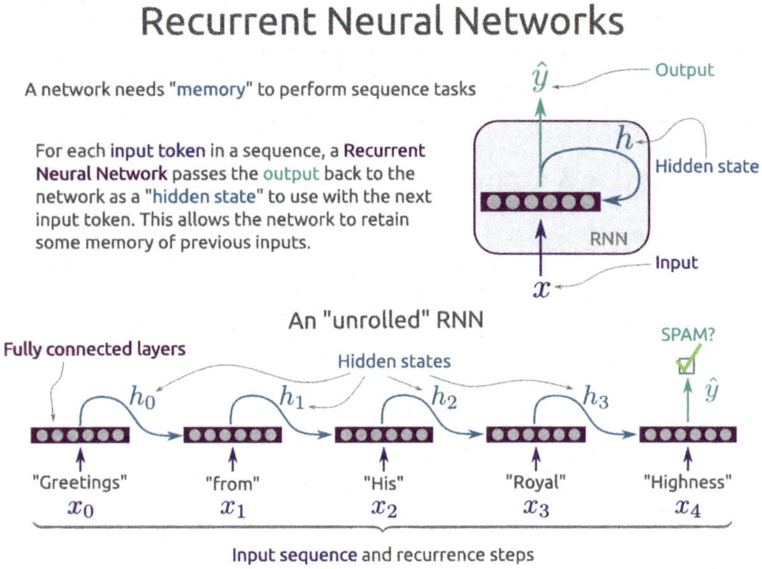

Recurrence in neural networks

The mechanics of a generic RNN are thus: For a sequence of N items, such as words, the network first looks at the first word, x_0, and produces two outputs, the first hidden state, h_0, and the first output, $\hat{y}_0$. The first output may or may not be used, depending on the task. The hidden state is passed back into the network for use with the second input, x_1. This process then repeats, until the entire sequence has been processed.

The hidden state captures information that can be passed through all steps of the sequence, providing some context for the task. Note that the network starts with no hidden state input, so an input of zero value is typically used.

More generally, the hidden state from step t can be represented as h_t and the output for step t represented as $\hat{y}_t$, with

$$h_t = \sigma(\mathbf{W}_{hh}^T h_{t-1} + \mathbf{W}_{xh}^T x_t)$$

$$\hat{y}_t = \mathbf{W}_{ht}^T h_t$$

where the **W**'s are learned weight matrices representing the the connections of a single layer[3] and σ is the layer's activation function. Each weight matrix takes a specific input to produce a specific output, as designated by the indices.

Backpropagation through time

As with other neural networks that we've seen, RNNs are trained via (versions of) gradient descent and backpropagation. The recurrent nature of the RNN requires backpropagation to effectively be performed not just once, but back through each step of a sequence. This is termed "backpropagation through time" or BPTT.

While the network's weights are "reused", each step in processing a sequence provides the network with additional information, so must be taken into account. You can think of backpropagation as taking place over the entire unfurled network. The loss function is a sum of the losses for each of the sequence step outputs (if there are more than one).

Creating context with embeddings

For language related tasks, words and how they are put together form the context needed to perform the task. The words (or usu-

[3]We're using the notation of a single layer network here for convenience. Real RNN's would likely have multiple hidden layers, but which would require more complicated notation to represent the iteration over the hidden layers.

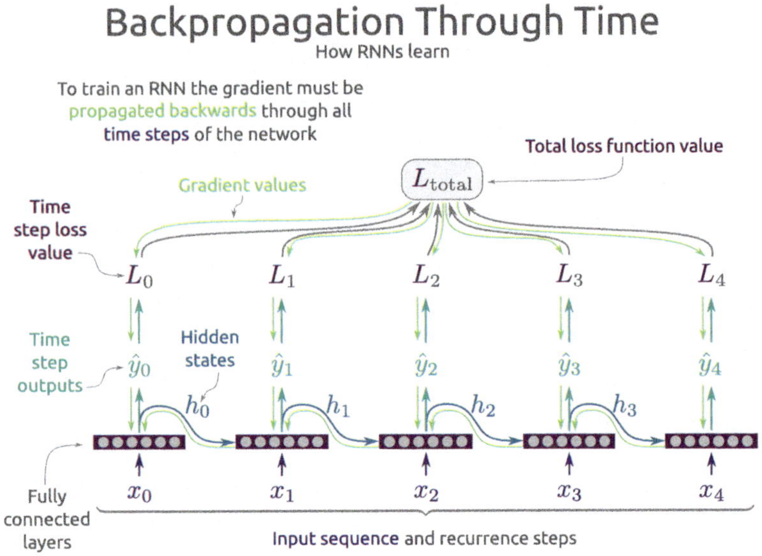

Backpropagation through time

ally word tokens[4]) can be thought of as the basic features for many language tasks. How can you use words as features though?

Computers need words to be represented numerically to make sense. The most generic way to do this is to have a dictionary, where each word (token) has a corresponding number associated with it. To input a word into a model, you look up the number. To understand the output of the model, you do a reverse lookup of the output number and find the corresponding word in the dictionary.

Semantically (or meaning-wise) words don't have an inherent ordering to them. There's no obvious reason why "cat" should have a higher or lower numerical value than "cup". Instead, the most

[4] A *token* is the smallest unit that a sequence is divided into for processing. For NLP, that might mean that a word is shortened to its stem or root or broken down into its stem and ending.

basic way to encode words is to use the process of *one-hot-encoding*. One-hot-encoding is when you create a vector of the same length as the total vocabulary size under consideration. If you have 1,000 words (or tokens) in your vocabulary, your input vector is an array with 1,000 entries (i.e. a vector in a 1,000 dimension space). For a given word, one (and only one) entry in its vector will have a value of 1 and all other entries will be zeroes (and are thus all orthogonal to one another).

One-hot-encoding is simple and useful, but it can be unwieldy for large vocabularies and it doesn't give us any sense of relationships between words. As language users, we know that words have inherent relationships. "Hot" is the opposite of "cold". A "boat" and a "ship" are closely related. In order to better capture the relationships, similarities, and differences between words, instead of one-hot-encodings, embeddings are used.

Embeddings

Embeddings are ways to represent data that are richer in information than simple numerical encodings, such as one-hot-encodings. Though definitions vary somewhat, I will use the definition that an embedding is a "learned, low dimensional representation of data that increases the information contained [for the task of interest], versus a simple enumeration".

What does that mean? It means that we want to be able to use the data itself to learn relationships and other information that is implicitly captured in the data. In Chapter 3 we saw how neural networks learn transformations to make the task of the network easier to perform. Embeddings are a version of this. The network learns how to transform the data in such a way that it is more useful than the raw data.

What does "low dimensional" mean? Low dimensional means using fewer dimensions (i.e. the length of the vector) to represent the data than you would need for one-hot-encoding. If your vocab-

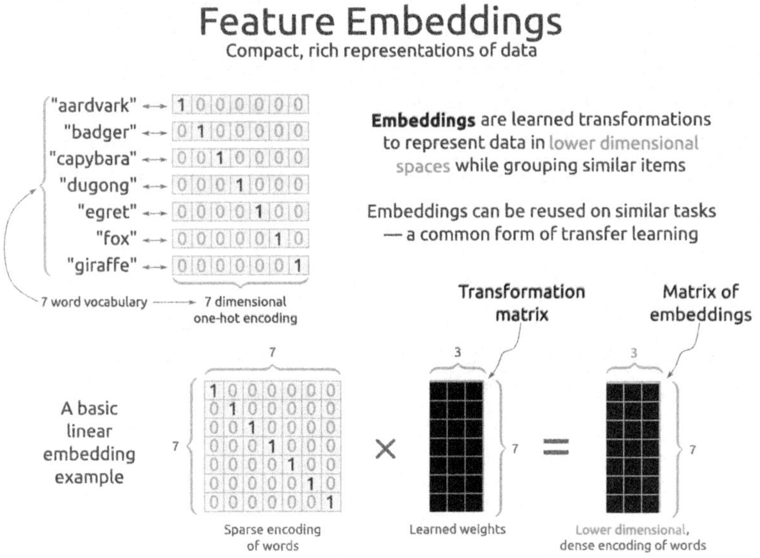

Embeddings

ulary had 1,000 words, instead of 1,000 dimension one-hot-encoded vectors, you might use an embedding vector of size 40. The size of the embedding vector (dimension) is a hyperparameter. Instead of a sparse, high dimensional encoding, the data is now in a dense, lower dimensional space.

If the embedding is created via a process that needs to extract relationship information from the words, it will necessarily create embedding vectors that exhibit these relationships.

Embeddings work for many types of data

Embeddings aren't just useful for NLP, but can be used with many types of data and tasks.

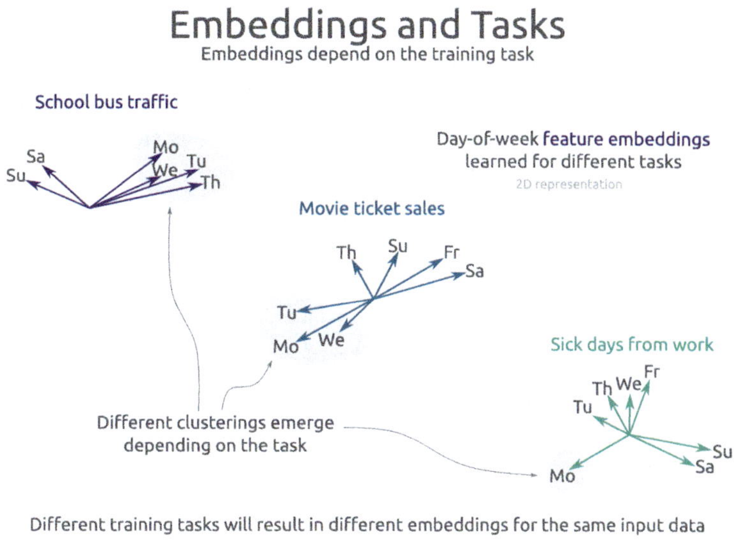

Embeddings of day-of-week data

> Imagine that you were using day-of-week as a feature in your data for some task, such as forecasting traffic. While day-of-week has a natural ordering and you could capture that with a simple encoding of 1 through 7, that may not capture as much information as it could. Is Friday equally (dis)similar to Saturday as it is (dis)similar to Thursday? Depending on the task, Friday may be much more like Thursday or much more like Saturday. For example, in the US, schools will be similarly busy on Thursdays and Fridays, whereas movie theaters will be similarly busy on Friday and Saturday evenings.
>
> By creating an embedding of the day-of-week in, say, 3 dimensions, you may be able to better capture the relationships of the day-of-week for your specific task.

Embeddings are very useful in several contexts. Because you

can compute the (dis)similarity of two embedding vectors with each other, you can use embeddings for things like search and ranking (especially when embedding things like images, websites, or products). One of the most common ways to compare two embedding vectors is with *cosine similarity.* Cosine similarity is simply the cosine of the angle between two vectors and ranges from -1 to +1. Vectors that have a smaller angle between them have a higher cosine similarity.

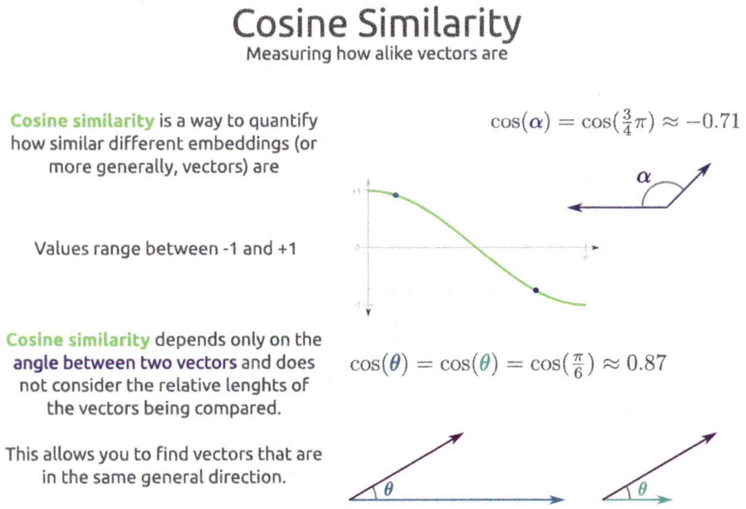

Cosine similarity

It's easy to think about embedding vectors for similar data being close to each other and thus having a small angle between them, but of course most embeddings are in high enough dimensions, that you can not visualize what this looks like.

The most straight forward way that embeddings are created

is by creating a "transformation matrix"[5], which, when multiplied with the one-hot-encoded vector, encodes (or projects) the one-hot-encoded vector into the embedding space, i.e. you go from your 1,000 dimension one-hot-encoded vector to the 40 dimensional embedding vector. In a typical neural network, this is the equivalent of adding an additional layer to the network.

For word related tasks, there are a few historically important methods for creating word embeddings, including word2vec and GloVe[6]. We will look in more detail at word2vec.

Word2vec

Word2vec is a method for creating word embeddings. It was introduced by Tomas Mikolov et al, in 2013[7]. It was notable, because the embeddings can be easily reused as feature transfer and the relationships in the embeddings often are inline with human intuition about word relationships.

Famously, the inventors showed that you could perform meaningful "embedding math", whereby basic math on the embedding vectors can produce results such as "brother" - "man" + "woman" = "sister" (i.e. if you subtract the "man" vector from the "brother" vector and then add the "woman" vector to that, the closest embedding vector to this result is the vector for "sister"). The conclusion being that word2vec (and similar) embeddings are able to represent words in ways that capture semantic meaning interpretable by humans.

---

[5]This matrix is known by many names, including "projection matrix", "embedding lookup table", "feature embedding matrix", and "input mapping matrix".

[6]Jeffrey Pennington, Richard Socher, and Christopher D. Manning, "GloVe: Global Vectors for Word Representation", 2014. `https://nlp.stanford.edu/projects/glove/`

[7]Tomas Mikolov, Kai Chen, Greg Corrado, and Jeffrey Dean, "Efficient Estimation of Word Representations in Vector Space", 2013. `https://arxiv.org/abs/1301.3781`

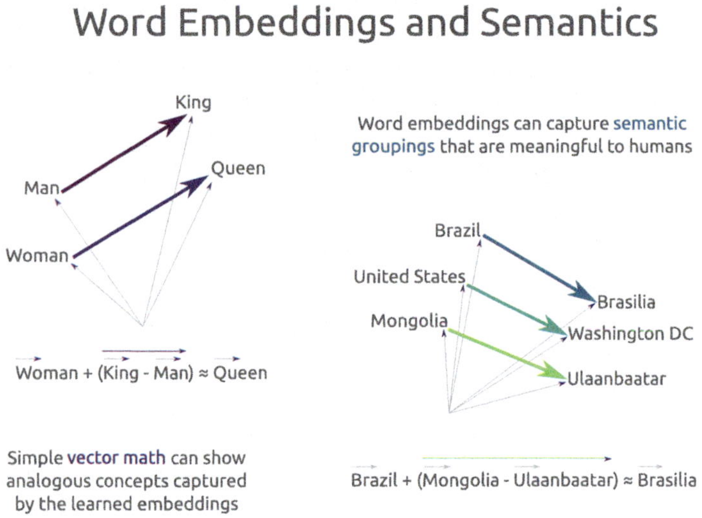

Word embeddings and semantics

Word2vec works by training a shallow neural network[8] to solve a self-supervised learning task. To do this, a large set of text data is used as a training set and an appropriate task is chosen. There are a number of training tasks that are used, but we will discuss building a model to solve the task called "skipgrams with negative samples".

Skipgrams are sequences of text where one word is the *context* word and the words surrounding it are blanked out. The goal of the task is to guess the words most likely found around the context word – sort of the opposite of "fill in the blank". The number of surrounding words is called the "window size" and is set by the user. It's essentially a multi-label classification problem. For a given context word input, the network needs to predict all of the

[8]A "shallow" neural network is simply a network with only one hidden layer.

correct labels, i.e. the surrounding words. By moving through the training text and choosing each word as the context word and the surrounding words as the target labels, you provide the model with a large set of self-supervised training data.

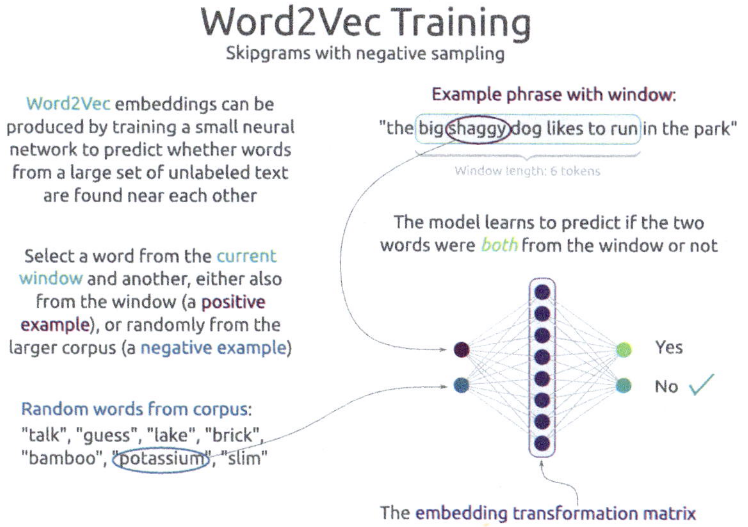

Word2Vec training

This can work, but one of the main problems with this task formulation is that the possible number of labels is essentially the size of the vocabulary, which can be very large. The creators of word2vec realized that they could simplify this task by turning it into a binary task instead. The new task is to simply classify whether two words, the context word and another word, would be found in the same sequence window. The window itself only provides positive samples for this task (i.e. words that *are* found in the same window as the context word), so you also need to generate negative sample words (i.e. words that would not be found in the same window). To do this, you can randomly select words from the full vocabulary of the text.

With negative samples, the task then becomes to train the model to classify pairs of words as coming from the window or not. For the model to perform well on this task, the network needs to transform the input one-hot-encoded words into a lower dimension space, such that words that would be found together in windows would be also near each other in this embedding space. Words not found together within the same window should not be near each other in the embedding space. This "nearness" (or similarity) is calculated within the network by performing an inner (or dot) product of the embedding vectors, which is closely related to cosine similarity.

Architectures for sequential tasks

RNNs are a powerful concept, but the generic version described above is not used much in practice. One of the main issues that they face is their inability to retain long-term memory, which is needed for many tasks. This stems from the fact that the longer an input sequence is, the more iterations of the network are needed. A very deep network like that makes the chance of the gradient vanishing greater, as there are even more terms in the backpropagation product[9]. Remedies, such as better choices of activation functions (e.g. ReLUs) and better parameter initialization, can help.

Researchers have developed modified RNN architectures to address the issue of memory. We'll take a look at two that have been important: LSTMs and GRUs. While LSTMs were developed some 20 years before GRUs, we'll discuss GRUs first, as they are simpler.

---

[9]See Chp. 3 for a refresher on vanishing gradients and remedies. Gradient explosion can also occur, but that can be dealt with via gradient clipping above a threshold value.

Gated recurrent units

Gated recurrent units (GRUs) were introduced in 2014 by Kyunghyun Cho et al. The main concepts they use beyond a basic RNN are the "candidate hidden state", noted as $\tilde{h}$, and a mechanism to update, mix, or forget the previous hidden state, called "gates".

There are two gate functions:

- The "reset gate" determines how much of the previous hidden state should be used to create the new candidate hidden state.
- The "update gate" determines how much of the new candidate state it should use and how much of the previous hidden state it should use as the new hidden state.

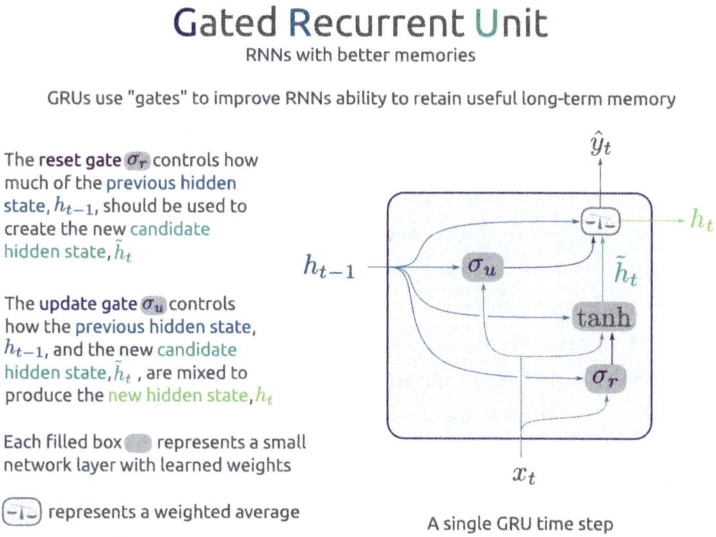

Gated recurrent units

Mathematically, the gate is a sigmoid function, producing a

value between 0 and 1 (but most of the time very close to either 0 or 1). To "gate" another value, the output of the sigmoid is simply multiplied by the other value, e.g. the previous hidden state.

Like most other aspects of neural networks, the amount of gating is learned. The inputs of the gate function are multiplied by weights. These weights are learned along with the other parameters within the network. The gate is essentially a small fully-connected network within the larger GRU network.

The GRU network is somewhat complicated, because it applies its gates more than once within the gated unit. After the reset and update gate values are calculated, the reset gate value is applied to the previous hidden state, h_{t-1}, which is then input, along with the step input, x_t, to the activation function that creates the candidate hidden state, $\tilde{h}$.

The update gate value is then applied to the candidate hidden state and its compliment (1 - the update gate value) is applied to the previous hidden state, i.e. a weighted average of the new candidate hidden state and the previous hidden state. This potentially allows hidden states to pass through the recurrent network with little to no change, if that "long-term" hidden state is important for the task. Conversely, this allows networks to forget hidden states that are not useful.

Because a hidden state can potentially pass a long way through the network with little to no change, this reduces the risk of vanishing gradients, as the gradient term from "early" in the network can effectively be the product of many fewer (very small) values.

Long short-term memory

GRUs are actually a simplification of LSTMs, a modified RNN architecture introduced by Sepp Hochrieter and Jürgen Schmidhuber in 1995. LSTMs also allow RNNs to have selective longer-term memory and overcome the vanishing gradient problem. Like GRUs, they use gates and candidate hidden states.

The main differences are that LSTMs use an additional "output gate" and pass two separate hidden states between each recurrence, one is the activation function output (used in RNNs and GRUs as the hidden state) and the other is a gated version of the activation function output, i.e. the same value, but multiplied by a factor between 0 and 1.

Similar to GRUs these gates and the resulting learned hidden state management allow LSTMs to be more robust and better at predicting the solutions to sequential data tasks. These differences make the LSTM generally more powerful than GRUs, but also somewhat more complicated.

Given the success that LSTMs have had and their (relatively) long history, the LSTM has been one fo the most influential deep learning architectures.

Attention

The modifications to generic RNNs introduced by LSTMs and GRUs allow for much higher performance, but they are not without limitations or issues. Due to the sequential nature of recurrence, RNNs cannot be parallelized for training, making them poorly suited for very large scale training sets. Additionally, for the same reason, they are limited to sequences of only tens or hundreds of tokens.

"Attention" is an alternative method of considering context in sequential tasks. It was introduced in 2014 by Dzmitry Bahdanau et al and has been a highly influential technique. Instead of focusing on the hidden state passed from the most recent recurrence, as in an RNN, attention determines how much weight should be given to each previous step[10]. Attention is especially relevant in sequence to sequence tasks, such as language translation or document summarization.

---

[10]RNNs can run forward, backward, or in both directions. For tasks such as language translation, bidirectional RNNs are advantageous, because the order of words varies in different languages and looking ahead can be important.

For producing translations, summaries, or similar tasks, the higher level architecture that is typically used is called an encoder-decoder architecture. One RNN encodes the input sequence into a hidden state. Another RNN decodes that hidden state into an output sequence.

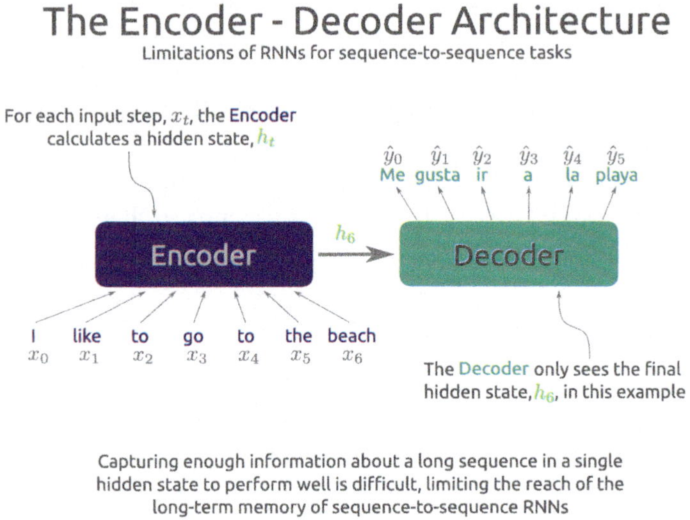

Limits of the encoder-decoder RNN architecture

An encoder-decoder architecture with RNNs, as described so far, means that longer sequences have more issues, as it's harder to maintain a rich representation of a long sequence. Attention attempts to solve this by allowing the decoder part of the model to focus on different parts of the input sequence for different steps of the output sequence.

Sentences commonly have important information spread across them, requiring a model to either memorize the entire sequence or selectively pay attention to those important parts. For example

> "**The dog**, that lived two houses away and liked to patrol the neighborhood, **started barking**."

> "**Der Hund**, der zwei Häuser weiter wohnte und gerne in der Nachbarschaft patrouillierte, **fing an zu bellen**."

> "**El perro**, que vivía a dos casas de distancia y le gustaba patrullar el vecindario, **comenzó a ladrar**."

All have phrases between the subject and verb clauses. Only looking at the words immediately preceding the verb phrase might cause the subject to be lost.

Attention is a quantification of how much the current output sequence step should consider different input sequence steps. It's calculated by looking at the hidden state of the previous output sequence step, which we'll denote as S_{t-1}, and the hidden states of the input sequence steps, e.g. $h_{t-3}, h_{t-2}, \cdots, h_{t+3}$.

The attention value is calculated by learning the weights for a small neural network that takes the previous state, S_{t-1}, and an input hidden state, e.g. h_{t-2}, and outputs an attention value, $\alpha_{t,t-2}$. This is calculated for each input hidden state. The sum of attention values, $\Sigma_i \alpha_{t,t_i}$ is normalized to equal one. The attention values are then used to created a weighted sum of the input sequence hidden states used to calculate the current sequence output, S_t.

With attention, models are able to better focus on what parts of the input are important to the output. While attention was very useful in the context it was developed, it went on to be even more important with the further development of *self-attention* and *transformers*.

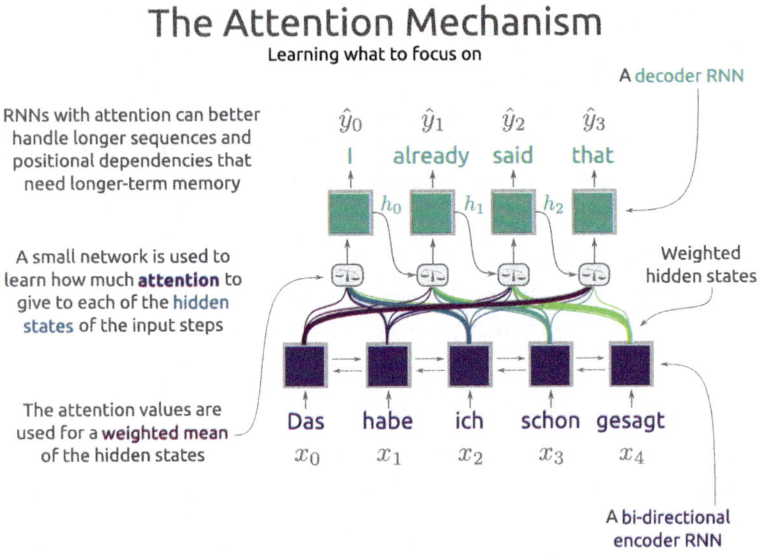

The attention mechanism

Transformers

In 2017, Vaswani et al published a groundbreaking paper titled "Attention Is All You Need"[11], which introduced the Transformer architecture. The Transformer is a novel network architecture that doesn't use recurrence as its main mechanism, allowing for parallelization of training and longer term memory. Paralellization is important, because it means that it can be trained on extremely large datasets by using many machines (i.e. GPUs or similar) at the same time. It is not an RNN, CNN, or fully connected NN, but instead uses only attention mechanisms as its main mechanism of computation.

[11] Vaswani et al, "Attention Is All You Need", 2017, `https://arxiv.org/abs/1706.03762`

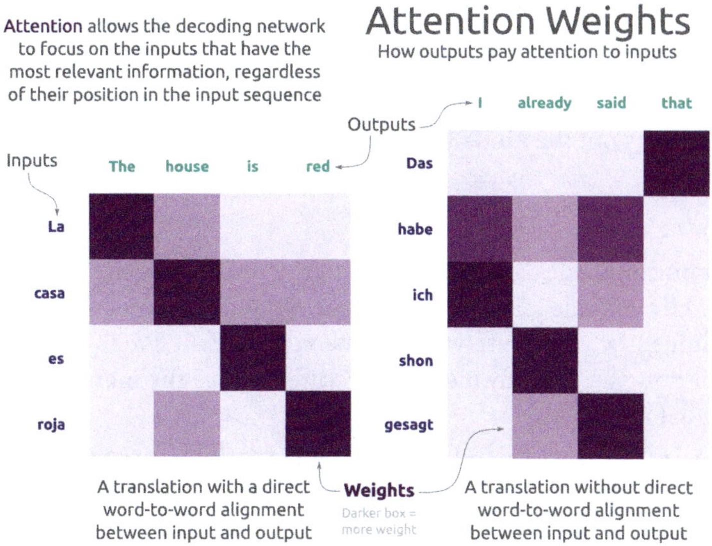

Attention weights

The Transformer architecture kicked off a slew of development that produced an improvement in NLP and sequence task performance, not unlike the spark that AlexNet lit for computer vision tasks. And like CNNs and ImageNet, Transformers (along with a couple of other architectures[12]) enabled much more powerful transfer learning for NLP and sequence tasks than previous approaches.

Much higher performance transfer learning was achieved by using "contextual embeddings", rather than just word-level embeddings. The fundamental issue with word-level embedding is that a single word can have multiple, often very different, meanings, but word embedding only allows for a single meaning for a given word

[12]Other important non-transformer models for NLP-related transfer learning included ELMo and ULMFit.

(without elaborate hacks).

Consider the word "kind" in the sentence

> He was not the **kind** of person that people described as **kind**.

Having only one embedding for "kind" would either end up capturing only one meaning of "kind" or ending up as some average of meanings, which likely wouldn't be very useful. We'll see in a little bit how the Transformer architecture enables the more useful contextual embedding.

The Transformer is a relatively complex architecture with several parts. We will cover the most important parts and innovations, while leaving some details out. Please see the resources section at the end of the chapter if you'd like to dive deeper.

The high-level architecture the Transformer

The Transformer architecture consists of two main parts: an encoder and a decoder. This makes it well suited for sequence to sequence tasks, such as language translation and summarization.

Both the encoder and decoder parts make use of two important innovations: so-called "self-attention" and "multi-head attention". Self-attention is a way for the network to determine which parts of the sequence are important in a way that is somewhat analogous to how a convolutional layer learns to extract or filter parts of interest of an image. Multi-head attention is essentially doing this in parallel, analogous to having multiple channels in a convolutional block in a CNN. Like a channel in a convolutional block, each attention head learns different features in the sequence appropriate for the task at hand.

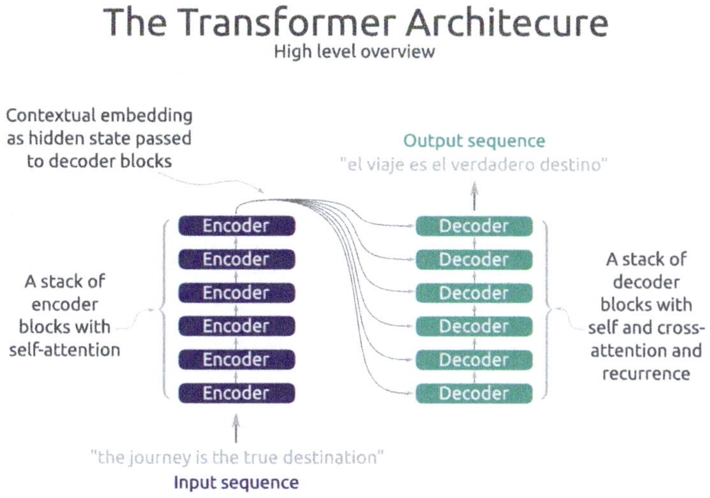

The Transformer architecture

Self-attention

We previously discussed "attention" in the context of an encoder-decoder RNN architecture. In that case the decoder is trying to determine which of the encoder hidden states to pay more attention to for a specific output step. With self-attention, the network is trying to figure out which parts of the input to pay more attention to for all output steps as a single process. In this sense, the network is comparing the current input to itself to understand which parts should get more emphasis. It can be used in an encoder-decoder architecture or just an encoder or decoder.

Self-attention is accomplished by having the network "ask" itself about what is important at each step in the network where self-attention is applied. This is formulated as having the network create "queries" from the layer's input tokens. The network then compares each query to a set of "keys" generated for the same inputs of the layer. Finally, there are "values" created from the inputs that correspond to each key. If a query and the key are similar, than the value for that key is given more weight (i.e. attention).

You can think of the "query", "key", and "value" arrays as learned embeddings. The query-key comparison is a search in the embedding space and more weight is given to values that are better matches.

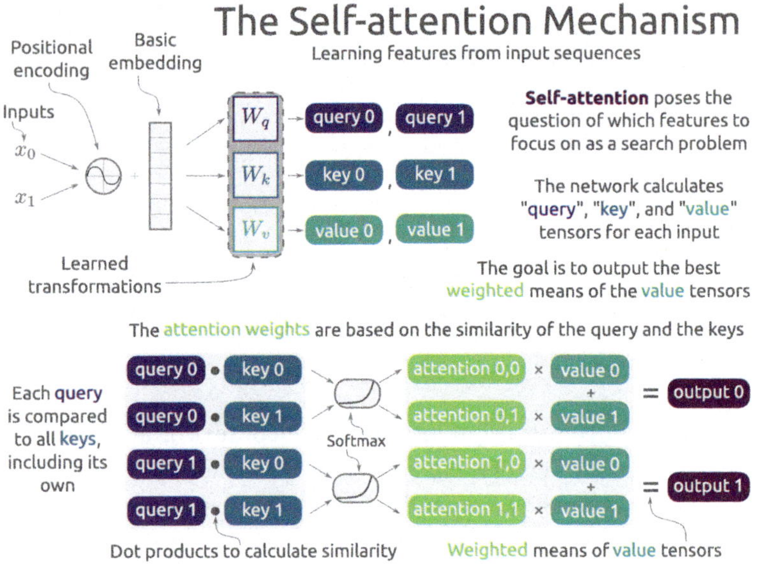

The self-attention mechanism

Mathematically these are matrix operations, where the input array (or *tensor*) for a given word in the input sequence is multiplied by the query matrix, Q, the key matrix, K, and the value matrix, V, to produce the query, key, and value arrays, respectively. The values in these matrices are model parameters that are learned during training. The similarity of the resulting key and value arrays are calculated by taking the inner product, producing a single weight. Each query is compared with all keys, including the query's own key.

The weight values for each word in the input sequence are calculated this way and then they are divided by a scaling factor and squished with a softmax[13] function. For each query array, all the value arrays are multiplied by there corresponding weights. These scaled arrays are then summed, producing the attention weighted outputs for each input of the self-attention layer (i.e. if there are four input tokens, there will be four outputs).

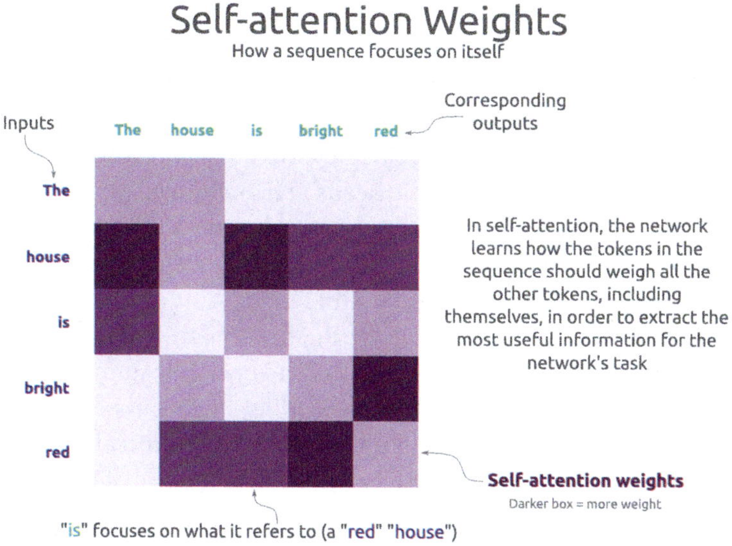

Self-attention weights

Although there is clear structure and intent to this design of

---

[13]Refer back to Chapter 4 for details on the softmax function.

queries, keys, and values, you can also think of this as a blackbox feature extractor that the network learns, similar to a convolutional block. Ultimately, self-attention is trying to take n inputs and return n outputs that are the best weighted sum of some embedded version of those inputs. The "best" weighted sum being whatever will best help the network solve its task.

The encoder and decoder blocks

The encoder in the Transformer consists of a self-attention block, a feed-forward network block, residual connections on both of them, and a few helper functions. The self-attention blocks are actually *multi-head* attention blocks: several self-attention layers in parallel, each with it's own query, key, and value mechanism. This allows the block to focus on different parts of the input sequence and learn different ways to represent what's in the sequence. This is similar to multiple channels in a convolutional layer in a CNN – each channel or attention head can perform its own feature extraction. The outputs of the multiple heads are combined.

Also analogous to CNNs having multiple convolutional blocks, the Transformer has several encoder blocks in series. Each of them having a chance to extract more useful features.

The input to the initial layer of the first block is different, in that an embedding layer is applied and a special position encoding operation is performed. The embedding layer puts the input sequence elements (i.e. words) into the right dimension array. The positional encoding allows the network to have a sense of where in the sequence each input is.

The decoder is very similar to the encoder side, but with a few differences to allow it to produce sequences as output. The input sequence to the decoder is actually from itself – it's a recurrent network that feeds its own output back in to build up a sequence. Because the decoder is producing the output sequentially, it uses a mask to ignore the "future" items in the sequence (even though

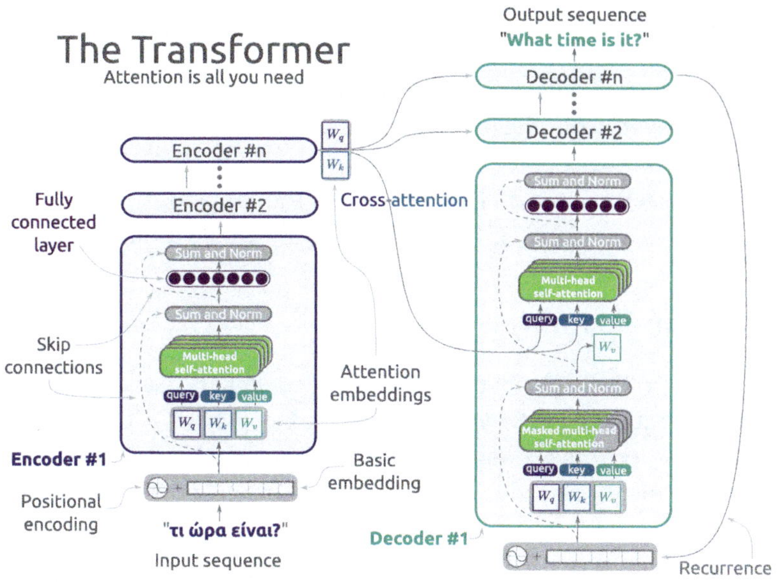

The Transformer architecture

they don't yet exist). This helps to reduce errors. This "masked attention" block comes before the (multi-head) self attention layer. The self-attention layer and feed-forward layers are the same architecture as in the encoder, except that the inputs are coming from two places: the masked attention block and the output of the encoder block. The output of the encoder is transformed into a new set of queries and keys for the decoder block's self-attention layer. This special mixing of inputs is called "cross-attention".

Applications and Transformer based architectures

The transformer architecture has become the building block for several other network architectures, which are the ones used in practice. These models have proven very useful for almost all of the language tasks mentioned in the first part of this chapter. While the original Transformer is an encoder-decoder architecture aimed at sequence to sequence (i.e. many-to-many) tasks, many of the models derived from the Transformer use only either the encoder or decoder block.

In this section we will look at some of the Transformer-derived architectures and some of the tasks that they are being used for.

Transformers for NLP

One of the first transformer based models to get wide use was BERT, a model created in 2018 by researchers at Google AI[14]. It was able to set many records on language related benchmarks. BERT stands for "Bidirectional Encoder Representations from Transformers". It only used the encoder section of the Transformer architecture, because that's all it needed for the tasks it was designed to perform.

BERT uses a Transformer encoder based architecture to learn contextual word embedding. Each word in an input sentence is embedded incorporating information from the sentence that it's found in. To accomplish this, BERT is trained on two self-supervised tasks: filling in the blanks in sentences and classifying whether two input sentences are consecutive. These pre-training tasks require that BERT implicitly learn language structure and semantic meaning.

After being pre-trained on a large unlabeled data set, BERT can then be used to produce contextual word embedding for other models or can be fine-tuned on a labeled data set for many-to-one tasks, such as spam classification, sentiment analysis, etc. Freely available pre-trained weights mean that BERT can be used to power all sorts of new tasks, in a way similar to many computer vision models.

[14] Jacob Devlin, Ming-Wei Chang, Kenton Lee, and Kristina Toutanova, BERT: Pre-training of Deep Bidirectional Transformers for Language Understanding, 2018. `https://arxiv.org/abs/1810.04805`

Another notable model is GPT-3, published in 2020 by Open AI. GPT-3 is a language generation model (or large language model, LLM) based on the Transformer decoder. The pre-training task is next word prediction. Like other Transformer based models and modern LLMs, GPT-3 was trained on a huge set of training data for this self-supervised task. GPT-3 had about 100 times as many parameters as its immediate predecessor, GPT-2, and a much stronger performance. GPT-3 was the first LLM to catch wide public interest, as its generated text seemed to cross a threshold of human-level language plausibility.

In addition to BERT and GPT-3, there have been many more Transfomer-based models, such as RoBERTa, T-5, and OPT. They've grown in network size (i.e. number of parameters), training data size, and performance level. Relative to CNNs, most of these language-related models require much more training data and hardware resources. With hundreds of billions, or even trillions of parameters, training these large models from scratch has become effectively impossible for individuals or small organizations[15]. This has increased the importance of transfer learning for NLP.

Transformers beyond NLP

Transformer-based architectures can be used for other sequence-based tasks besides natural language. In fact, if a task can be formulated as a sequence-based task, Transformers can be applied, including image and time-series related tasks.

The Vision Transformer (ViT) is a BERT-like Transformer applied to computer vision tasks. It was introduced in 2020 by Google Brain[16]. It works by treating an image as a sequence of image patches. ViT works by learning which parts of the input image to

[15] At least as of the time of writing this book.

[16] Alexey Dosovitskiy et al, "An Image is Worth 16x16 Words: Transformers for Image Recognition at Scale", 2020, `https://arxiv.org/abs/2010.11929`

pay attention to. Instead of words, the input tokens are the image patches embedded into an array of tokens.

ViT and many related models have become popular for computer vision tasks, such as image classification, object detection, and segmentation. As of 2022, they are not yet the standard for computer vision problems, but they are competitive for many tasks.

Summary and resources

In the last chapter I claimed that computer vision is arguably the most successful application of deep learning in the past decade. It's very possible that the era of Transformer-based models means that NLP related applications are nearly as well addressed by deep learning. Even further, there is speculation that both computer vision and NLP tasks will soon all be performed on the same architectures: Transformer-based architectures[17].

While Transformers are currently the "new, hot thing" in NLP and sequential data tasks, understanding RNNs is still important, as they are still in use for certain tasks and were a major step in the evolution of neural network-based approaches to solving sequence-based tasks.

Below are some resources to help you improve you understanding of NLP, sequential tasks, and deep learning techniques for them.

Courses

- Andrew Ng's course on Sequence Models[18] on Coursera.
- Fast.ai's Practical Deep Learning for Coders[19], with Sylvain Gugger and Jeremy Howard.

---

[17] AI researcher Andrej Karpathy argued that there is an ongoing consolidation across all task domains towards Transformers in this Twitter thread: `https://twitter.com/karpathy/status/1468370605229547522`.

[18] `https://www.coursera.org/learn/nlp-sequence-models`

[19] `https://course.fast.ai/`

- Stanford's CS224n: Natural Language Processing with Deep Learning[20] by Chris Manning
- The NYU Deep Learning[21] course by Yann LeCun & Alfredo Canziani.
- MIT 6.S191 Introduction to Deep Learning[22], with Alexander Amini and Ava Soleimany.

Books

- "Zefs Guide to Natural Language Processing[23]", by Roy Keyes. Coming 2023.
- "Natural Language Processing with Transformers: Building Language Applications with Hugging Face" by Lewis Tunstall, Leandro von Werra, and Thomas Wolf. O'Reilly Publishing.
- "Machine Learning with PyTorch and Scikit-Learn: Develop machine learning and deep learning models with Python" by Sebastian Raschka, Yuxi (Hayden) Liu, and Vahid Mirjalili. Packt Publishing.

Other Online Resources

- The Unreasonable Effectiveness of Recurrent Neural Networks[24] by Andrej Karpathy.
- Visualizing A Neural Machine Translation Model (Mechanics of Seq2seq Models With Attention)[25] by Jay Alammar.
- The Illustrated Transformer[26] by Jay Alammar.
- Transformers from Scratch[27] by Brandon Rohrer.

---

[20] `https://web.stanford.edu/class/cs224n/`

[21] `https://atcold.github.io/NYU-DLSP21/`

[22] `http://introtodeeplearning.com/`

[23] `https://zefsguides.com`

[24] `https://karpathy.github.io/2015/05/21/rnn-effectiveness/`

[25] `https://jalammar.github.io/visualizing-neural-machine-translation-mechanics-of-seq2seq-models-with-attention/`

[26] `https://jalammar.github.io/illustrated-transformer/`

[27] `https://e2eml.school/transformers.html`

Chapter 7

Advanced techniques and practical considerations

Beyond the task specific architectures introduced in the last two chapters, there are many techniques and practical considerations important for people working with deep learning to be familiar with. In this chapter we will discuss some advanced architectures, common techniques, and some of the problems and tasks that need to be solved in order to make deep learning useful in the real world.

Because this chapter will be wide-ranging, I will include pointers to resources in the sections that they are associated with, rather than at the end of the chapter.

Combining vision and language

So far we have discussed image related tasks separately from NLP, text, and sequential tasks. But as you may imagine, there are many tasks that require or benefit from combining multiple types of data. These are referred to as "multi-modal" tasks and models.

In this section we will take a high-level look at a few multi-modal techniques that combine both image and text data.

Image captioning

One of the tasks that was mentioned in the last chapter was that of image captioning, which is simply adding a descriptive sentence or phrase to go along with an image. While we have considered both image and text-based tasks, how we could combine these may not be obvious. As is turns out, the key is embeddings.

In Chapter 6 we discussed how embeddings are a way to take data and create new richer, typically lower dimensional, representations of it by learning a transformation (i.e. projecting the data into a lower dimensional space). We can then use the embeddings as inputs to other systems or networks.

For image captioning, a common way to achieve this is to use a (pre-trained) CNN to produce embeddings as the initial hidden state input to an RNN or Transformer decoder model – an example of feature transfer.

The decoder is trained with a labeled dataset of images with accompanying captions.

Joint embeddings

Embeddings are "abstract" representations of data as vectors. Unlike images, sentences, or other human-interpretable types of data, they are just arrays of numbers optimized to contain as much useful information as possible. While that may seem intangible, it also presents us with the opportunity to create relationships between different data types for solving tasks such as search and recommendations.

"Joint embeddings" are the outcome of training models to produce embeddings of different data types (e.g. images and text) such the embeddings of similar concepts are similar across data types. In other words, an image of an oak tree would have an embedding vector very close in embedding space with the embedding of the words "oak tree". To do this, the different encoding networks must

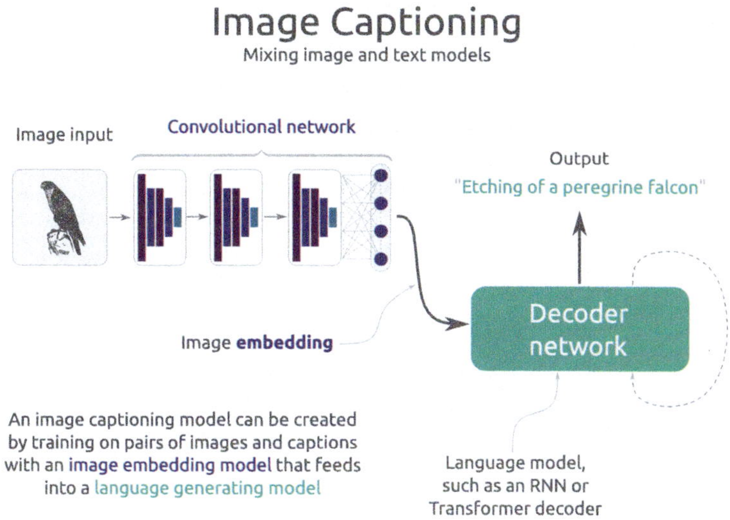

An image captioning model

be trained to put these separate embeddings together.

A well known example of joint embeddings is the CLIP[1] model (a.k.a. Contrastive Language Image Pre-training). CLIP uses an image encoder, such as a CNN or Vision Transformer, and a Transformer text encoder to produce both image and text embeddings for image-caption pairs. CLIP has the twin goals of embedding pairs having very high similarity scores, while embeddings not from the same pair in the training dataset having as low similarity as possible.

A joint embedding model such as CLIP can be used for either text-based image search or image-based caption search. A new search term can be run through the respective encoder and the im-

[1] `https://openai.com/blog/clip/`

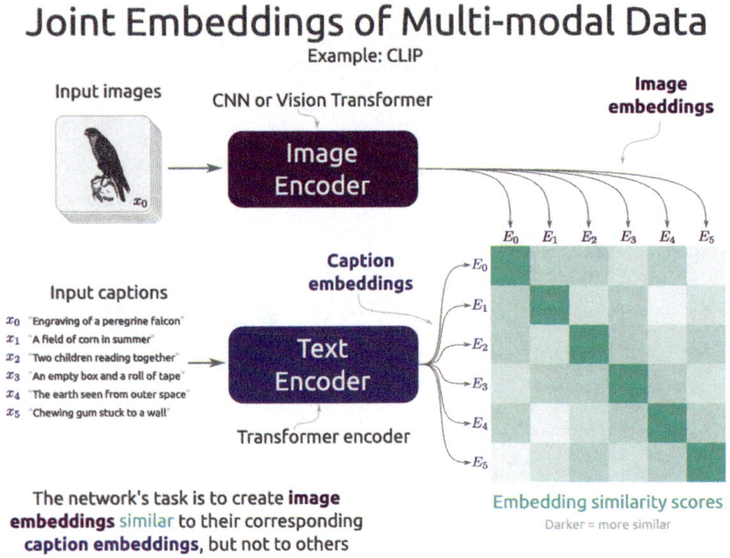

Joint embeddings of multi-modal data

ages (or captions) in the dataset with the most similar embeddings can be returned. This type of embedding comparison is the basis for many kinds of search and recommendation systems.

Diffusion models

Diffusion is an image generation technique that has taken the deep learning world by storm in 2022[2]. Similar to GANs discussed in Chapter 5, diffusion learns to construct images from scratch. Unlike GANs, diffusion uses a multi-step generation process that seems to be more robust during training and ultimately seems to produce better results.

---

[2]I'm very curious to see how this innovation ages.

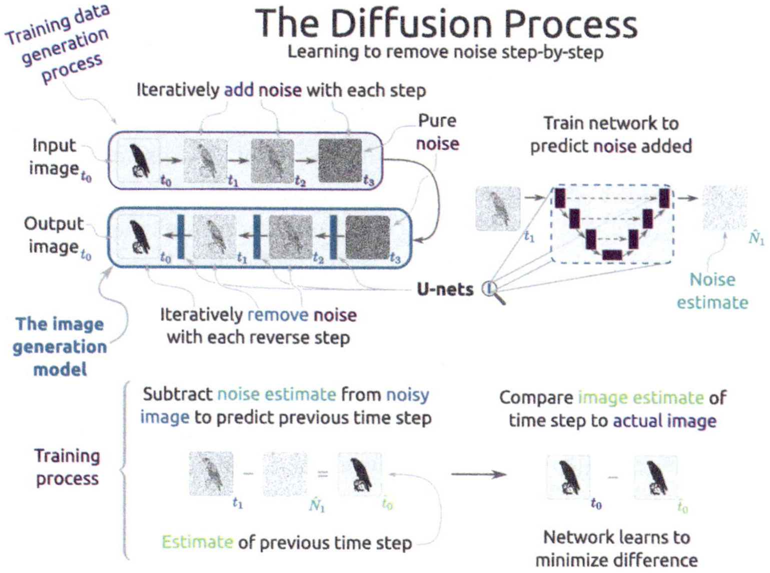

The diffusion process

The name "diffusion" refers to the cumulative addition of Gaussian noise to images in a way that is mathematically identical to particles diffusing over time. The technique is inspired by thermodynamics.

The diffusion process is actually about learning to remove noise from images. The training process starts by generating training data by adding noise to a "clean" input image over many steps until the image is pure noise. Next the network is trained to try to remove noise step-by-step until the original image is restored.

To remove noise, a U-net[3] tries to estimate the noise in the image at each step. For a given step, the noise estimate is then removed from the image and the resulting clean image estimate is

[3]See Chapter 5 to review U-nets.

compared with the known clean image for that step. This comparison is the basis for performing gradient descent and optimizing the U-net parameters.

Typically diffusion uses tens or even hundreds of steps, as it appears to be easier to remove a small amount of noise at a time. Once a diffusion network is trained, it can start with pure noise and iteratively remove noise until a clear image is produced. A GAN works in essentially the same way, except that it needs to make the jump from pure noise to a final image in a single step, which has proven to be less robust than doing it many steps.

Stable Diffusion

While you can take an image of pure noise and run it through a diffusion network to get an image, it turns out that you can both direct the image generation process and produce a clearer image by using a "prompt" to guide to the process.

A prompt is simply another piece of data, such as a text description or another image, such as a sketch. Unsurprisingly, to input that prompt into the network, you're gonna need an embedding.

In 2022 several diffusion models with text and/or image-based prompts where announced, including DALL-E 2[4] from OpenAI, Imagen[5] from Google Research, and Stable Diffusion[6] from Stability AI. These models work by training a diffusion model to incorporate a prompt (embedding) to "condition" the output of the model. The U-net in the diffusion model learns to change its output based on the input prompt.

Arguably, Stable Diffusion has made the biggest impact of these models, as its code was released publicly for all to use. Stable Diffusion works in a slightly different way than the other models mentioned, as it performs diffusion in the "latent space" (essentially the

[4] `https://openai.com/dall-e-2/`

[5] `https://imagen.research.google/`

[6] `https://stability.ai/blog/stable-diffusion-announcement`

same as the embedding space) rather than in "pixel space". We will look at Stable Diffusion as an example, but overall it is very similar to the other popular diffusion-based image generation models.

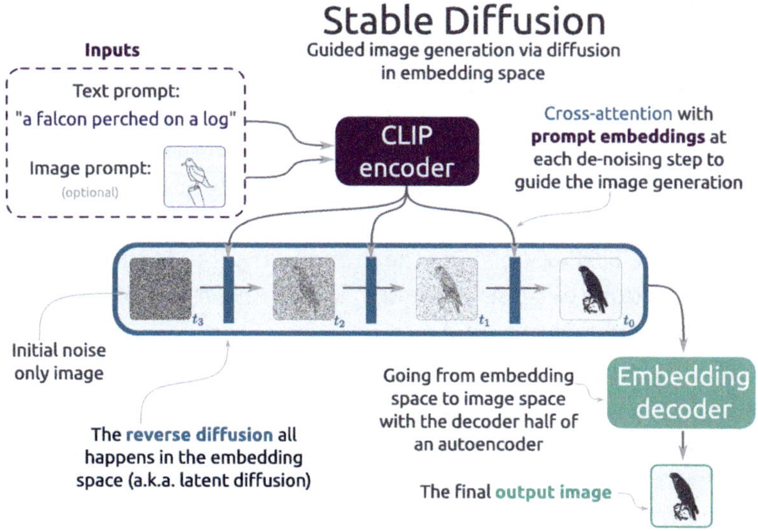

Stable diffusion

The generic method that Stable Diffusion is based on is called a "latent diffusion model" (LDM[7]). Instead of removing noise directly in pixel space, latent diffusion learns to remove noise in the embedding (or latent) space. The process is essentially the same, except that the input image used to generate the training data is first embedded with a CNN or similar image encoder. The U-net is trained on data in the embedding space as well. Only once the de-noising of the image is complete, is the image transformed back into pixel

[7]High-Resolution Image Synthesis with Latent Diffusion Models[8], Robin Rombach et al, CVPR, 2022, `https://arxiv.org/abs/2112.10752`

space. Latent diffusion is basically standard diffusion sandwiched inside an image autoencoder.

The reason for working in embedding space is primarily because the embeddings are much smaller than the raw images, making the process computationally more efficient.

The prompts, whether text or image, are embedded with a CLIP-style model. During inference[9], the U-net is run with both the prompt as an input and with a null prompt as an input (i.e. an array of zeros). The output of these are compared, with the idea that the embedding space vector corresponding to the difference between the prompted output and the prompt-less output points in the direction of the best de-noising. The prompted output is then moved in that direction (by some amount). This is essentially a trick and is termed "classifier-free guidance".

To enable the prompt to guide the de-noising, cross-attention is included in the U-net to allow it to better incorporate the prompt embedding information.

Taken all together, these steps have proven to be very powerful in creating highly realistic and novel images when trained on large, high quality datasets.

Self-supervised learning

Most of this book has been concerned with supervised learning, where a model is trained on a dataset where the answers (i.e. the labels) are known. We also touched on unsupervised training, where a model learns to look for groupings of similar data within a dataset without the use of labels to guide it.

A third paradigm for training models is *self-supervised* learning, which is in someways a mix of supervised and unsupervised

[9]In ML, "inference" just means making predictions with the model after it's trained.

learning[10]. Acquiring unlabeled data is often much easier and/or cheaper than acquiring or curating labeled data. Self-supervised learning is a solution to this problem[11].

In self-supervised learning, a supervised learning task is chosen, such that the data labels can be deterministically derived from the data itself. This often means "scrambling" the original data in some way, with the original state of the data serving as the label. Examples include reordering the letters of a word, where the correct order is the label, learning to colorize a photo, where the input to the process is the original image, but in grayscale, and the target is the original, color image.

Once a model is trained on this self-supervised task, the model can be used as the basis for transfer learning – either as feature transfer (e.g. using the outputs as embeddings) or using a labeled dataset to finetune the model for a related task.

In previous chapters we have seen several examples of self-supervised learning already, including autoencoders and GANs (Chapter 5), word and text embedding techniques, including word2vec and Transformers (Chapter 6), and diffusion techniques in this chapter.

Image-based techniques

Text tends to present relatively straight-forward self-supervised techniques, such as fill-in-the-blanks or predicting whether two sentences are next to each other. Image data can be used in similar ways.

Common image-based self-supervised tasks include slicing images into tiles and having the model predict the correct ordering, removing a section of the image and having the model fill in the hole,

[10] Historically this was most often grouped with unsupervised learning before the term "self-supervised learning" gained popularity.

[11] It can be argued that it is also more similar to much learning done by naturally intelligent systems.

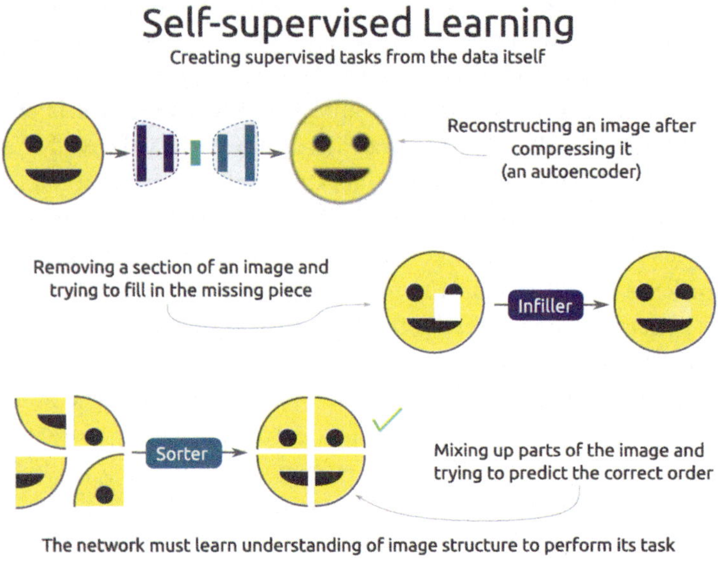

Self-supervised learning for images

flipping and/or rotating the image and having the model predict the correct transformation, or any techniques were reconstructing the input is the goal.

All of these tasks are designed such that achieving the explicit task would require the model to implicitly learn a strong understanding of image composition and how to consider context within the image.

Contrastive learning

Constrastive learning is a self-supervised approach, where the goal is to train the model to learn to produce the "correct" output well and to be bad at producing the "incorrect" output. That means that the model needs to learn from both "positive" and "negative" labels.

In typical supervised learning this is inherent in the data, as the training set will (hopefully!) have examples from other classes, so the model needs to be able to predict all different classes.

We have seen examples of this style of self-supervised learning already: word2vec skipgrams with negative sampling and CLIP. Both present the scenario where you could just train on the "positive" examples, but instead are greatly enhanced by also trying to predict against non-positive examples as well. In word2vec the task boils down to whether two words are from the same window of words. By also asking the model to compare against words from outside the given window, the "contrastive" aspect is added. In CLIP the model needs to force the paired image and caption embeddings to be similar, but also the image embeddings should be dissimilar to the caption embeddings of other images.

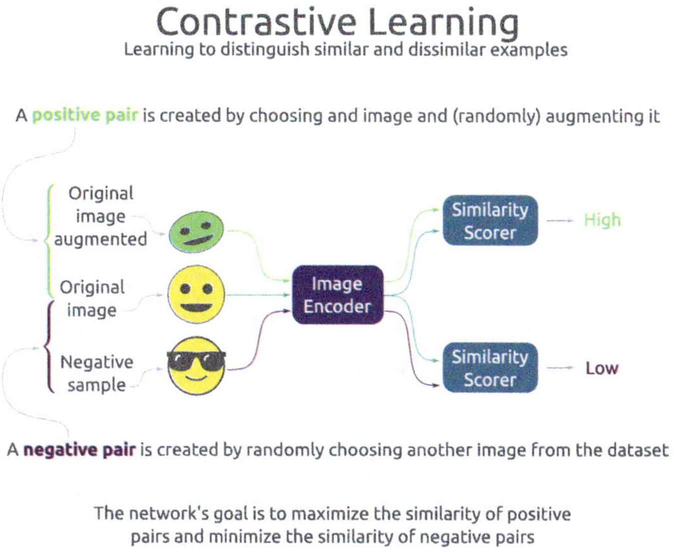

Contrastive learning

As with these examples, the "contrastive" or "negative" examples for training are just randomly sampled from the larger dataset. Because this is not human labeled, there's no guarantee that every contrastive sample will actually be contrastive (i.e. you might randomly select a word that happened to also be in the same window or a caption from an image that was similar). The goal is that *statistically* there will be enough contrastive examples to produce a strong model.

As with most topics in this book, we've only scratched the surface of contrastive learning, but the key idea is there: learning better representations of the data to improve your ability to perform subsequent, related tasks (a.k.a. the "downstream" task).

Some survey articles on self-supervised and contrastive learning techniques:

- Longlong Jing and Yingli Tian, "Self-supervised Visual Feature Learning with Deep Neural Networks: A Survey[12]", 2019
- Xiao Liu et al, "Self-supervised Learning: Generative or Contrastive[13]", 2020

Math topics related to deep learning

Neural networks are implemented in computer code and to build them (from scratch) you need to know how to program computers. But, to understand how they work in detail, you also need to understand certain math topics. In this section we will briefly discuss the most important mathematical topics for deep learning.

Linear algebra

Most of the mechanics of neural networks are described in terms or arrays (or tensors), matrices (also tensors), and mathematical oper-

[12]`https://arxiv.org/abs/1902.06162`
[13]`https://arxiv.org/abs/2006.08218`

ations you can perform on them. You can often describe most aspects of neural architectures compactly as matrix multiplications. The weights and biases learned during training are entries in arrays and matrices.

All of this is in the domain of linear algebra. Understanding linear algebra gives you the tools to understand how neural networks function from a different perspective and especially to understand technical descriptions in academic papers, etc. Most of the core parts of neural networks are some combination of operations between vectors and matrices in different configurations.

Some resources for beginners in linear algebra:

- Khan Academy's Intro to Linear Algebra[14]
- MIT's Intro to Linear Algebra[15] with Gilbert Strang
- Linear Algebra for Everyone[16] by Gilbert Strang, Wellesley-Cambridge Press, ISBN 978-1-7331466-3-0, September 2020

Statistics and probability

Statistics and probability are at the heart of all machine learning. Machine learning is about recognizing and approximating the general pattern in the data. Instead of memorizing the training data, a well-trained model will produce good predictions, because it has learned the general patterns in the data.

Understanding concepts of distributions, summary statistics, probability, etc, are key to really understanding and working with machine learning. Not only does statistics underlie how neural networks work, but also how you evaluate the performance of ML models. Performance metrics are statistical measures of how good

[14] `https://www.khanacademy.org/math/linear-algebra`

[15] `https://ocw.mit.edu/courses/18-06-linear-algebra-spring-2010/`

[16] `https://math.mit.edu/~gs/everyone/`

a model's predictions are. Additionally, to compare two models, you may want to employ statistical hypothesis testing.

Some resources for beginners in statistics and probability:

- Khan Academy's Statistics and Probability[17]
- OpenIntro Statistics[18]
- Udacity's (free) Intro to Statistics[19]

Differential calculus

The "learning" part of deep learning is finding the right parameter values for the network to best perform at its task. As we saw in Chapter 3, gradient descent is the key method used to optimize a network's parameters. At the heart of gradient descent is calculus.

While most deep learning practitioners are not solving many integrals or calculating derivatives by hand, calculus is the mathematics of ML parameter optimization. Understanding derivatives (in particular partial derivatives) is important for understanding how and why much of the optimization process works in ML and DL, in particular. It's often said that backpropagation is really "just the chain rule" for derivatives.

Some resources for beginners in calculus:

- Khan Academy's Calculus 1[20]
- Khan Academy's Multivariable Calculus[21]

Machine learning engineering

So far we have mostly discussed the concepts behind deep learning in a relatively abstract way. Real world usage of deep learning mod-

[17] `https://www.khanacademy.org/math/statistics-probability`
[18] `https://www.openintro.org/book/os/`
[19] `https://www.udacity.com/course/intro-to-statistics--st101`
[20] `https://www.khanacademy.org/math/calculus-1`
[21] `https://www.khanacademy.org/math/multivariable-calculus`

els involves much more than just the ability to train a "good" model. Creating a machine learning-based system that is running "in production" involves many moving parts and design considerations. Building these systems falls under "machine learning engineering".

Deep learning libraries

Rather than writing model code from scratch, deep learning practitioners use libraries designed specifically for creating deep neural networks. Libraries such as PyTorch[22] and TensorFlow[23] center on optimized tensor operations, but also have many built-in conveniences for creating neural networks.

Graphical processing units and specialized hardware

As mentioned in Chapter 4, graphical processing units (GPUs) have played a critical role in the rise of deep learning. Most neural network operations can be formulated as matrix operations, which GPUs where specifically designed for.

GPUs are typically used in the training stage of neural networks and sometimes in the inference stage, depending on the specific application. The cost of purchasing or renting GPUs (in the cloud) can be the limiting factor in the training of models.

Other specialized hardware, such as Google's Tensor Processing Unit (TPU), has been developed to support neural networks. Deep learning libraries, such as PyTorch and TensorFlow typically come with support for (multiple) GPUs and TPUs built-in.

Machine learning systems

Models are usually just a single, small part of real-world machine learning systems. A typical ML system that runs "in production"

[22] `https://pytorch.org`

[23] `https://www.tensorflow.org/`

will include:

- Tools and infrastructure for the sourcing, storing, labeling, managing, and maintaining of training data.
- Tools and infrastructure for training, experimenting, and managing and tracking experiments.
- Tools and infrastructure for serving predictions and managing production models.
- Tools and infrastructure for monitoring predictions and production models.

A real-world ML system needs to be designed to not only make good predictions, but also be maintainable and upgradable. The first version of the model is almost never the last version. Even with a "perfect model" running today, the goal of the system is likely to evolve and the data itself is likely to shift over time.

Building and/or orchestrating all of these pieces fall under the umbrellas of machine learning engineering (MLE) and machine learning operations (MLOps).

Some other considerations for the design of ML systems include how and where models will be deployed (and sometimes trained). If a model needs to run in real time on a device, for example, the system will need to be optimized to be small and fast, sometimes for specific "edge" hardware. There is an array of techniques for making networks smaller and less resource intensive, such as "pruning", "quantization", and "distillation".

Further factors when designing models and ML systems involve considerations such as privacy, regulatory compliance, fairness, and other constraints or objectives.

Some resources on machine learning engineering, systems, and operations:

- D. Sculley et al, "Machine Learning: The High Interest Credit Card of Technical Debt", 2014, `https://research.google/pubs/pub43146/`

- Shreya Shankar et al, "Operationalizing Machine Learning: An Interview Study", 2022, `https://arxiv.org/abs/2209.09125`

Wrapping up

The history of neural networks is a long and filled with booms and busts. By all indications, though, "this time is different". The deep learning era kicked off in the early 2010s has arguably brought more innovation and, crucially, more value than any of the previous eras of neural networks or artificial intelligence.

This book's goal is to give the reader a conceptual overview of deep learning, touching on the most import concepts and topics, as well as diving deeper into some of the most important technical aspects. Hopefully the book succeeded in helping you gain a more complete understanding of how everything fits together and provided you with some good jumping off points for further learning.

The breakneck pace of innovation in deep learning means that it's more difficult than ever to keep up with the latest techniques and technologies[24]. Hopefully this book has made it easier for you to at least understand the big picture of deep learning and how the latest techniques fit in, if not actually understand the technical aspects of those techniques.

I am sure that we are in for many more interesting developments. Please be on the lookout for additional books in the Zefs Guides series!

---

[24]During the writing of this book in the first half of 2022 I had only intended to include diffusion as a bullet point in a list, but due to the sudden explosion interest in diffusion-based models in the second half of 2022 I decided it warranted inclusion. Things move quickly!

Printed by Amazon Italia Logistica S.r.l.
Torrazza Piemonte (TO), Italy

58916468R00097